The Rightful Place of Science:

Science, Values, and Democracy

The 2016 Descartes Lectures

The Rightful Place of Science:
Science, Values, and Democracy

The 2016 Descartes Lectures

Heather Douglas

Foreword by
Sir Peter Gluckman

Edited by
Ted Richards

Consortium for Science, Policy & Outcomes
Tempe, AZ, and Washington, DC

THE RIGHTFUL PLACE OF SCIENCE:
Science, Values, and Democracy
The 2016 Descartes Lectures

Consortium for Science, Policy & Outcomes
Arizona State University

Printed in Charleston, South Carolina.

The Rightful Place of Science series explores the complex interactions among science, technology, politics, and the human condition.

For information on The Rightful Place of Science series,
write to: Consortium for Science, Policy & Outcomes
PO Box 875603, Tempe, AZ 85287-5603
Or visit: http://www.cspo.org

Model citation for this volume:

Douglas, H. 2021. *The Rightful Place of Science: Science, Values, and Democracy: The 2016 Descartes Lectures*. Tempe, AZ: Consortium for Science, Policy & Outcomes.

ISBN: 099958779X

ISBN-13: 978-0999587799

FIRST EDITION, AUGUST 2021

CONTENTS

FOREWORD

This volume contains three edited lectures originally delivered in 2016 by Heather Douglas. Each is followed by two commentaries that illustrate a depth of critical discourse welcome not only for scholars, but also for practitioners whose work is at the interface of the many domains explored in this book. That Heather Douglas has given lectures in a series named for René Descartes seems highly appropriate. After all, Descartes has for centuries influenced Western thinking about values and empiricism. Douglas herself has influenced much current thinking about the practical nature of the interface between science and values. These nine well-linked contributions will be of essential value not only to scholars of the philosophy of science, but also to practitioners at the interfaces between science and policy and science and society.

Science should not be defined as a form of truth, but rather as a set of processes for the making and testing of empirical knowledge claims in an ongoing and cumulative way. Done well, it is communal, self-reflexive, and, thus, self-correcting. The issue that emerges in the context of science's interface with democracy has to do with its claim of general epistemic authority. More specifically: On what basis is such authority earned or granted? As this volume discusses, such authority can only exist when well supported by values, principles, and behaviors of scientists and the scientific community. But how the policy community, political actors, and society then act is inextricably intertwined with the community of science. It is complex, messy, and evolving. But a greater understanding of these interfaces is essential if democracy is to thrive, and if science is to play its essential role in addressing the challenges of sustainability, resilience, and human development.

Although there are many scholars who have interrogated the intersections of science with society, policymaking, and politics, too many of those insights have stayed buried within the confines of specific scholarly disciplines. Seldom can this work claim any obvious impact on the actual conduct of science or science advising or science communication. Even post-normal science (what Arthur Petersen would rebrand as "extra-normal" science in this volume), which seeks to describe and explore the boundaries from an experiential perspective, has had little traction within the relevant communities. Douglas, however, makes a strong case that for science and democracy to serve each other well, an understanding of the place of values and principles must be well embedded within the practitioner. In these lectures, as in her previous book, *Science, Policy, and the Value-Free Ideal*, she makes this argument cogently. Understanding when values are appropriately embedded within science and when not—and appreciating that the use of science is always a values-based choice—is key to ensuring that science appropriately supports democracy.

The questions addressed in these nine contributions encompass issues of immense importance, issues which COVID-19 has brought to the fore. How should we understand the interdependence between science and democracy? What are the responsibilities of scientists when they function as experts in society? How should the tensions between the principles of democratic scrutiny and transparency operate against the need for decision makers to have trusted relationships with experts? How, in an age of misinformation, should skepticism and denial be addressed? How can we enhance literacy about the processes that shape and define science?

The discussion could be further advanced by exploring the two distinct components of science advising to policymakers: *evidence synthesis*, given the plurality of inputs and perspectives that Douglas has already pointed out are necessary; and *evidence brokerage*, which is the careful and appropriate transfer of the knowledge from that synthesis to the policy and political process and to the public. These are distinct undertakings, often involving different actors and very different theoretical and practical considerations. Unless this distinction is made clear, the questions of accountability and transparency become difficult to parse. My concern is simply that the discussion these papers have initiated is not yet complete and hopefully this will be an area for further reflection.

Each of the chapters, both the lectures by Douglas and the commentaries by her interlocutors, opens up issues in a very accessible way, but importantly does not close them. There is much more to reflect upon and explore, particularly if the work is to transcend the theoretical and impact the practical, as much of Douglas' work has already done. This book is a valuable stimulus for scholars and practitioners alike. In the age of COVID-19, post-truth politics, and digitally accelerated misinformation, the importance of understanding and confronting these issues cannot be overstated.

Sir Peter Gluckman
President-Elect, International Science Council
Chair, International Network for Government Science Advice
Director, Koi Tū Centre for Informed Futures, University of Auckland
August 2020

INTRODUCTION: WHY SCIENCE, VALUES, AND DEMOCRACY?

Heather Douglas

Human beings are social creatures. We live together, learn together, and love together. This sociality is part of the success story of human beings—together, we can accomplish things that would be quite impossible individually. Science is one of the enterprises that requires sociality among humans and is one of the things that no individual can quite manage on their own. Scientists require other scientists to collaborate with, to read and discuss ideas with, even to compete with. Without other scientists, there would only be individual investigation, a dull and idiosyncratic affair without the grit of other minds against which to hone ideas.

Essential to science, at least to science done well, is discord and disagreement, the willingness to challenge each other's ideas, which makes us work better than we would alone. Such disagreement can arise from different experiences, different cultural backgrounds, and different social locations. The plurality of voices, rather than unanimity, is part of what makes science such a potent social institution and effective knowledge production system.

The plurality of voices needed in science must come from somewhere. That science is embedded in society, that it is not held fully apart or conceptualized as autonomous and isolated from the society in which it functions, proves crucial. We live in pluralist societies, with disagreements about values and ways of living, with divergent experiences, with local architectures and expectations. The plurality that exists in our human societies is a resource for doing science well.

But such pluralism is also a source of difficulty for governance. That we must collectively govern ourselves arises from our very sociality. We cannot each find our own patch of wilderness in which to live, like tigers roaming forests, particularly as the world's population continues to grow and stretch the earth's resources. Most of us have no desire for such an isolated existence, as much of what we find valuable in human endeavors (such as art, music, friendship, and science) requires the sociality of human existence. But we don't all agree. So how, in the face of the sociality of humanity and the plurality of humans, can we collectively make decisions on how to live?

There have been many experiments with different forms of government over the past centuries, but democracy has proven to be "the worst form of government, except for all those other forms that have been tried," to borrow a quip from Winston Churchill. We are perpetually frustrated and occasionally horrified by what democratic governance produces, but no system of government is as self-correcting (particularly in the face of the rampant potential for abuse) as democratic systems. Indeed, among political theorists, there are few advocates for non-democratic systems—only debates about which kind of democratic system should be pursued[1] and, of course, which economic system should accompany democratic governance.

Democracies encompass many different forms, but the central characteristic that creates an umbrella over them all is an institutional check on those who govern by those who are governed. There is an ongoing accountability built into the system, so that those who are granted power to make decisions on behalf of the citizenry can be peacefully recalled from that power, if the governed disagree with the direction of the government. This is usually minimally in the form of elections, in which citizens vote and thus have a definitive say in the governance process. Democracies usually also involve a range of other social structures, such as independent judiciaries, freedom of the press (to encourage reporting on and public deliberation over the actions of those in power), and public ability to openly question those in government. Different countries and cultures instantiate the required mechanisms, which produce public deliberation and recall power differently, but in all, the ongoing consent of the governed is the central source of legitimate power.

With this legitimate power, democratic societies create binding agreements about how we live. Democratic governments can legislate what can and cannot be sold in the marketplace, what is and is not going to be supported with the public purse, and what laws or regulations will bind public behaviors. The ability of democratic governments to shape so deeply how we live means we cannot stop paying attention to what our governments are doing. Being in a democracy means an ongoing commitment to engaging in the disputes and struggles of the society (not all of the time, but at least some of the time), as the perpetual experiment of democratic governance unfolds.

In this series of papers, I will not make any ontological claims about whether science or democracy has metaphysical primacy, or that one is foundational or causative of the other. I find such arguments generally suspect. Robert Merton argued in 1942 that democracies are always more conducive to science, but nondemocratic states have produced good science.[2] More recently, Timothy Ferris has argued that scientific endeavors produced democratic states in the eighteenth century.[3] But the history is more complex than in Ferris's narrative and the causal arrow from science to democracy is far from straight, as the most robust scientific cultures (e.g., the Royal Society in the United Kingdom) were late in coming to democracy.

Further, recent scholarship provides a much richer account of the development of democratic forms, discovering far more texture than the usual highlights of Athens and the American Revolution.[4] Rather than such essentialist stories, all I presume here is the fact that we want both science (for the discovery and ongoing testing of empirical knowledge claims) and democracy (for the governance of pluralist societies). That we want both is enough for getting on with, for it makes clear that we need to think carefully about how these two central institutions—science and democratic government—are to relate to each other within our human societies.

That there is a challenge here at all is clear from the differences between science and democratic governance in how they decide things. While both science and democratic governance value deliberation and debate, democratic governance requires closure of decisions, often before consensus is reached. We need to know what the socially binding rules of action will be, so that we can get on with our lives, and we cannot wait for everyone to agree. We often use voting to achieve such closure—for example, voting among our elected representatives or voting for a referendum or voting for the representatives themselves. The vote is a binding act and, through it, decisions that grip onto the whole of society are made, even if in the next legislative round they can be reopened.

In science, we rarely require such closure, especially before consensus is reached organically. To vote to end a discussion among scientists about what to think would be anathema in science, unless there were pressing external reasons to do so. This is part of the luxury of the space of inquiry in which science and other academic pursuits exist, that inquiry can go on indefinitely (resources permitting).[5] And when scientists speak in worried tones about democratizing science, it is the conflict between science and democratic governance with respect to practices that lurks in their minds, that somehow science could be reduced to voting on results.

Science, particularly scientific expertise, can also pose a threat to democracies. It would be devastating for democracies if scientific experts were given legislative authority outside of accountability mechanisms, if our democracies devolved into technocracies. In addition to the need for public officials to be aware of relevant science, we need experts to be kept informed by, and in touch with, the needs and interests of the public. Ultimately, we need socially binding decisions to remain democratically accountable.

Despite these sources for conflict, there are areas where science and democratic governance must interact. In areas of science funding (either from the public purse or in areas of moral concern), of the use of science in policy (how science informs democratic decision-making), and of the influence of science on society generally (its ability to "legislate" our lives through, for example, the shifting of public/private boundaries or the creation of new capacities), we must sort out how these two central aspects of human societies are to interrelate.

I will begin my investigation of this question with the examination of science and values. The pervasive (and often legitimate) influence of social and ethical values on science creates an avenue through which democratic politics might productively and legitimately interact with science (without devolving to the caricature of voting on scientific results). Once I provide an account of values in science, I will then turn to an examination of how scientists and elected officials should interact. How should those in power and those who investigate the world relate to each other? How should we understand the responsibilities and roles for those who would "speak truth to power"? Finally, I will turn to the democratic public and assess what we need to do to create robust public discourse about science, given its pervasive importance for both society and governance.

I originally gave these lectures as the René Descartes Lectures in September 2016 at Tilburg University in the Netherlands. I am grateful for the invitation from Jan Sprenger and Silvia Ivani and for their work organizing the event, for those who attended the lectures and the rich conference created around them, and to my commentators for providing such excellent reflections on the material. Thanks as well to Joyce Havstad and Ted Richards for reading and discussing early drafts with me. Although it has taken four years to finally get the lectures ready to publish, I have not substantially altered them, and they reflect faithfully what I said in 2016. The world has changed in many ways since these were given. We are in the midst of a global pandemic, for which scientific expertise has been absolutely essential, and the success of using such expertise can be assessed in weeks or months, well within the span of public memory. The disparities of the pandemic's impacts have also been palpably on display, fueling a long-overdue focus on social justice. Renewable energy sources have become cheaper than fossil fuels, showing us how policy and infrastructure are now the main impediments to effective action on climate change, still looming as a long-term threat. In the pressure cooker of these and other crises, the threats to both democratic governance and scientific practice have intensified. I continue to believe we need both science and democracy to maintain healthy, pluralist human communities, and hope these lectures help to that end.

Notes

[1] Political theorists debate the merits of representative democracy versus direct democracy versus deliberative democracy, as well as when we should use what form.

[2] Robert K. Merton, "Science and Democratic Social Structure," in *Social Theory and Social Structure* (New York, NY: The Free Press, 1968), 604.

[3] Timothy Ferris, *The Science of Liberty: Democracy, Reason, and the Laws of Nature* (New York, NY: HarperCollins Publishers, 2010).

[4] Benjamin Isakhan, "The Complex and Contested History of Democracy," in *The Edinburgh Companion to the History of Democracy*, Benjamin Iskhan and Stephen Stockwell, eds. (Edinburgh: Edinburgh University Press, 2012), 1.

[5] Scientists do sometimes have to vote on how to distribute resources for doing science, as it is a decision that requires timely closure.

Lecture 1

SCIENCE AND VALUES: THE PERVASIVE ENTANGLEMENT

Heather Douglas

Since the mid-twentieth century, insulating science from social and ethical values has been something of an obsession for philosophers of science.[1] Philosophers articulated, and then staunchly defended, a value-free ideal for science. This ideal did not insulate science completely from societal influence. Philosophers were willing to concede the "context of discovery" to the influence of values (which, in contemporary parlance, includes scientists deciding upon research projects and methodologies), but argued that the "context of justification" had no place for social values. This view was supported by three ideas: 1) that societal values can add no confirmatory weight to empirical claims (and that to think otherwise is to confuse "ought" claims with "is" claims); 2) that values distinctive to scientific theory choice could guide scientists when faced with inferential decisions (i.e., epistemic or cognitive values); and 3) that the authority of science in the public sphere rested on the separation and disentanglement of science from social and ethical values. This final presumption was bound up with hopes for science as a resource in public debates that could transcend divergent societal interests—that science could be a "value-neutral" resource in our democratic discourse.

In this lecture, I will argue that there is something to the first idea—that there is an important conceptual difference between normative and descriptive claims, although in practice they are both used to support each other. Yet because of their logical structures, normative claims cannot provide sole support for descriptive claims, and vice versa. I will argue that the second idea is crucially incomplete—that although there are distinctive epistemic values in science, they cannot decisively guide inference. And, finally, I will argue that the third idea is inadequate as well—that we need a more complex understanding of why we grant science general epistemic authority, with multiple bases supporting that authority.

Descriptive and Normative Claims

It is a standard presumption in philosophy that one cannot derive "is" claims from "ought" claims, nor can one derive "ought" claims from "is" claims. I can make arguments about how one ought to value science or how one ought to value democracy, but that does not mean that the people to whom I am making these arguments *do* value science or democracy. Similarly, I can describe the way the world is in great detail, but someone can always respond, "Yes, but that is not how it ought to be." The difference between "is" and "ought" claims seems crucial for giving us the space to imagine what a better world might be like, even in the face of an (often grim) accurate and detailed description. It also provides space for resisting the automaticity that can follow from a particular "ought" claim. That the world is not like it should be (in some people's eyes) may be a good thing in our view, and we might use that descriptive difference, or the projected costs of change that arise from a detailed description, to resist a normative appeal.

Philosophers debate whether the practical difference between these types of claims is grounded in some metaphysical difference in the nature of normativity. Is the true different from the good? Is the beautiful different from the just? I have no wish to wade into such debates, although it certainly seems plausible that the answer to both these questions is yes. The world does not seem so unified that all the normativities line up. For my purposes here, it suffices to note that making a whole set of descriptive claims (with nothing else) does not make an adequate argument for a normative claim; nor does a whole set of normative claims (with nothing else) make an adequate argument for a descriptive claim. Each kind of claim cannot serve as sole justification for the other. They simply don't interact that way.

They do, however, interact. For example, it is difficult to see how to make an argument about how the world ought to be (or, more pointedly, how we ought to act) without relying upon *some* descriptive claims about the way the world is. We need empirical information about what causes pain, for example, if we are to craft a world with less pain in it. Arguments about what we should do rely upon descriptions of what we can do, what is feasible, what is readily achievable, what comes at higher costs and what those costs are. That we need both kinds of claims in our arguments is indicative that there are in fact two kinds of claims.

Conversely, the question of whether one can make an argument for a descriptive claim without normative claims is a central concern in the current values in science debate, particularly as science is now a major source for our descriptive claims. At issue is whether descriptive arguments rely on (without being built wholly out of) normative claims. The decisions of what to study, how to study it, and when to say the study is completed (when the evidence is sufficient) suffuses normative presuppositions into our descriptive statements. They are not there on the surface (just as descriptive claims are not there on the surface of normative claims) and they do not suffice on their own for arguing for (or supporting) the descriptive claims, but they are part of the overall argument for, and process of, generating the descriptive claims.

But to say that normative claims have a role to play in generating descriptive claims is not to give away the difference altogether. When Carl Hempel argued back in the 1960s that normative claims can provide no confirmatory weight to a descriptive claim, he was right.[2] Saying that the world ought to be a particular way is not a good argument that the world is actually that way. In this form of argument, such normative expressions are more pious hopes than reasons for the accuracy of descriptions. And this is a gap in kind that we want to preserve. The world is often not how we want it to be, and keeping this difference is essential for being able to perceive and to say that.

With this conceptual distinction in place, we can now address the debate over values in science. Acknowledging that there is a conceptual difference between descriptive and normative claims, we can examine more closely how they might (and should) interact in producing science. (For those who are not convinced there is a difference between descriptive and normative claims, the value-free ideal for science doesn't make any sense. In the arguments that follow, I will presume a conceptual difference, but show a practical interdependence, between the two.)

Values in Science

Science is, of course, a human practice. And when we do science, we entangle values, including social and ethical values, in that science. The questions are how values interweave with science, whether it is legitimate and necessary, and ultimately what to do about it.

Critics of the value-free ideal for science initially pointed out how values (particularly social and ethical values) influence the practice and products of science, because science is performed by humans. Feminist philosophers of science showed how sexist values blinded scientists to alternative explanations of phenomena or directed the attention of scientists to some narrow subset of data, a fuller examination of which produced rather different interpretations and results. Examples from archaeology (explanations of how tool use developed), cellular biology (explanations of fertilization processes), and animal biology (explanations of duck genital morphology and mating behavior) demonstrate such influences of values on science in spades.[3] Feminists were quick to point out that problematic science of this sort was not obviously *bad* science (scientists were not making up data and not engaged in pseudoscientific practices immune from criticism and revision), but the limitations of it (and the value-influence on it) became obvious once better science was pursued.[4] Looking back on the cases critiqued by feminists, the science looks woefully inadequate and blinkered.

Several feminist philosophers proposed addressing these issues by focusing on the social nature of science.[5] Because science requires communities of scientists, in critical dialogue with each other, feminist scholars looked to the structure of those communities for answers. Improving the diversity of scientists, many argued, would improve the range of explanations pursued and the kinds of phenomena examined, bolstering the epistemic reliability of the sciences. And by having more diversity of participants in science, and more diversity of values through the participants, the value judgments and influences could be more readily spotted by someone in the scientific community rather than disappearing, invisible by virtue of universal acceptance among scientists. In addition, in such an agenda, the virtues of the just and the true could be aligned, as breaking down the barriers to participating in scientific research would be both fairer and produce more accurate science.[6]

While this is certainly a worthwhile approach to addressing many issues of justice in science and epistemically inadequate science, this approach does not take on the value-free ideal directly. One could argue that the reason for increasing diversity in science is to ferret out those hidden value presuppositions that were distorting the search for truth.[7] Once made clear, one could hope that the values could be removed from the scientific explanations. The called-for diversity in science could be made to serve the ultimate aim of a value-free science. What the feminist critiques showed (for some) is not a problem with the value-free ideal per se but with the past practices of science. The cases of sexist science were weak science, empirically feeble science, and the pursuit of new theories and evidence made science stronger. Stronger science could still aim to be value-free.

Another reason the value-free ideal remained mostly unscathed was that it was narrowly focused on when values need to be kept out. One could still argue that science should be value-free in its justifications, that regardless of how the theories and explanations of empirical phenomena were developed (and feminist critiques showed we needed to improve this process substantially), what mattered when making inferences in science—when deciding what the evidence said—is that scientists try to keep values out of *that* process, and just focus on the evidence at hand (perhaps bolstered by a sense that with diverse participants in science, the evidence at hand is the best available set). The value-free ideal was articulated as being about the moment of inference in science, of being about the practices of justification at one particular point. The idea was that if values were kept out at this point, it could serve as the pure fulcrum for later decisions, that science could be universal and authoritative if and only if values were not part of the justificatory inference. And indeed, the idea that values can offer no confirmatory weight to the pile of evidence, and that if they did we would be blurring the important difference between descriptive and normative claims, added further reason to support the value-free ideal. Scientists needed to make inferences (and justify those inferences) with no regard to social and ethical values, according to the value-free ideal. Maintaining this ideal was crucial to the authority of science, which rested on purity from societal influences at the point of inference.

To upend the value-free ideal, and its presumptions about the aim of purity and autonomy in science, one needs to tackle the ideal qua ideal at the moment of justification. This is the strength of the argument from inductive risk. It points to the inferential gap that can never be filled in an inductive argument, whenever the scientific claim does not follow deductively from the evidence (which in inductive, ampliative sciences it almost never does). A scientist always needs to decide, precisely at the point of inference crucial to the value-free ideal, whether the available evidence is enough for the claim at issue. This is a gap that can never be *filled*, but only stepped across. The scientist must decide whether stepping across the gap is acceptable. The scientist can narrow the gap further with probability statements or error bars to hedge the claim, but the gap is never eliminated.

How is a scientist to decide that the available evidence is enough? That the gap is worth stepping across? That a claim is worth accepting? Some have suggested that epistemic and/or cognitive values can do this. It is time to examine whether there are "canons of inference" that can fulfill this role.

Epistemic and Cognitive Values: What Guidance?

When Isaac Levi suggested in 1960 that there were "canons of inference" that guided decisions of acceptance in science, and that these were sufficient for theory assessment in science, he helped to put in place a crucial piece of the value-free ideal.[8] There has been voluminous work on what became known as "epistemic values" (for some, "cognitive values") in science. Some of the work has focused on particular attributes (e.g., What is the value of simplicity? Does prediction matter more than accommodation? What constitutes a good explanation?),[9] and discussions initially described a collective soup of values that scientists held.[10] More recent work has involved unpacking nuance among the values considered constitutive of science.[11]

It has helped enormously to consider what these values are good for. Instead of merely noting their pervasive importance in science (historically and currently), one could attend to differences in why particular values might be central to science. For example, successful prediction and explanation are values that organize the evidence in relation to theory, and as such help to structure how we assess the strength of the available evidence.[12] Precision in successful explanation and prediction similarly helps assess how strong the evidence is—if precise theories explain or predict precise evidence, we think the evidential support is so much the stronger for the theory. Theories that successfully predict or explain a broad scope of evidence (across a range of phenomena), or theories that successfully predict or explain complex phenomena with simpler theoretical apparatus, also are judged to be supported more strongly by the evidence than competitors without these virtues. These kinds of values are properly epistemic, as they help us judge how good a theory is at this moment, and how strong the currently available evidence is.

Note that while these virtues are very helpful in assessing the strength of the available evidence, they are mute on whether the available evidence is *strong enough* to warrant acceptance by scientists. Such epistemic values do not speak to this question at all.

Other traditionally constitutive values in science are more future oriented, and direct our attention to the promise of a theory in the future. These values, such as broad scope over potential (but as yet ungathered) evidence, fecundity in producing predictions (as yet untested), and explanatory power (as yet uninstantiated), are suggestive of the general fruitfulness of a theory. But such future fruitfulness is only a reason to keep working on a theory, to use that potential fecundity to explore the world further, to accept it as a basis for further research, not to accept it generally for other decision-making. I have called these values "cognitive values," because their presence means that a theory will be easier to work with going forward, and thus they have a pragmatic research value for scientists.[13] They are not epistemic, as they do not indicate the general reliability of a theory—they do not tell us that a theory is well supported and likely to produce accurate predictions. They do, however, indicate good research bets. Thus, neither the set of epistemic nor the set of cognitive values can tell us when we have enough evidence. They simply do other jobs.

There are two sources of trouble here for seeing the normative entanglement of science and social values. The first is local: that many of the cognitive values have the same name as the epistemic values, and thus are readily conflated. Predictive power could be a name for past successes (and thus be epistemic) or could be a name for future fecundity (and thus be cognitive). The same goes for explanatory power or scope or even precision and simplicity. That there is a sense of these values that is directed to past success in grappling with and organizing actual evidence (an epistemic sense) and that there is a sense of these values that speaks of the future promise of a theory (a cognitive sense) confuses things. It also makes it seems as though the general list of such values is indeed sufficient for science—for what else do scientists qua scientists need but to assess the strength of evidence and to decide upon the future promise of potential research questions?

But such a conception of scientific practice neglects that we want something else from scientists; that indeed, science is not just pursued for scientists alone. We need to know what to think about the world right now, and not just to know which theories are promising for future research. And we need to know more than how strong the evidence is for a particular theory—we need to know whether it is strong enough to use for deciding what to do in the wider world beyond the endeavors of scientific researchers. The inductive gap remains, despite the utility of epistemic and cognitive values, and we have to know what to do about it. Should it be stepped across or not? Even with probability statements or error bars, does the available evidence support the claim enough? Epistemic values can help assess how strong the evidence is; cognitive values can help assess where to place bets for future research. But for the assessment of evidential sufficiency in the moment, we need to look beyond epistemic and cognitive values.

The Necessity of Social and Ethical Values in Science

How do social and ethical values help with this inductive gap? While they can't fill it, they are crucial for deciding when the evidence available (the strength of which is assessed using epistemic values) is strong enough. Strong enough for what? What is this assessment of sufficiency? How does a scientist decide that the inductive gap is acceptably small enough to step across? It is here, at this question, that philosophers and scientists must stop looking at the purely internal practices of science and answer this question with respect to a full understanding of science as it operates *within* societies, rather than *isolated from* societies. When scientists decide the evidence is strong enough, they are deciding not just for themselves, but for anyone who wants to rely upon science for guiding decisions in the broader world. For that, the internal practices and values of science are not sufficient.

Social and ethical values, however, do help with this decision. They help by considering the consequences of getting it wrong, of assessing what happens if it was a mistake to step across the inductive gap (i.e., to accept a claim), or what happens if we fail to step across the inductive gap when we should. In doing so, such values help us assess whether the gap is small enough to take the chance. If making a mistake means only minor harms, we may be ready to step across it with *some* good evidence. If making a mistake means major harms, particularly to vulnerable populations or crucial resources, we should change our standards accordingly. Social and ethical values weigh these risks and harms, and provide reasons for why the evidence may be sufficient in some cases and not in others.

The difficulty is that there are risks of error in all directions. There are risks of error in prematurely making a claim; there are risks of error in failing to make a claim soon enough; and there are risks of error in saying nothing while we wait for more evidence. There is no perfectly safe space in which to stand. Neither science nor logic can assure us of safety—indeed nothing can. There are no guarantees. What this examination of science, values, and inference can give us is not assurances of success, but assurances that we are doing the best we can—and what that best consists of. Doing our best in science requires the involvement of social and ethical values in the decision that evidence is sufficient.

There are alternatives to involving social and ethical values in evidential sufficiency assessments. We could simply toss a coin when deciding whether to accept or reject a claim. But this would be arbitrary, and thus irresponsible to the authoritative weight that science has in society. And we would still need to decide when the evidence was enough to warrant the coin toss! We could also set standards internal to science: What are the risks to scientific researchers and to the practice of science of accepting or rejecting a claim? But that is also arbitrary—arbitrarily insular: Why should impacts on scientists and research be the only impacts that count? Note that this too would still involve ethical values (some of the impacts on scientists would surely be ethically weighty), but we would be considering only scientists. Why should we do that? With science taking place within a broader society, why should only scientists count in making these decisions? We could ask that scientists never step across inductive gaps, but merely tell us the evidence and how strong they think it is. The practical difficulties of this are insurmountable. As I have argued, the moment of inference is not the only place where inductive risk considerations arise.[14] In addition, we would have to learn how to examine the evidence ourselves, as scientists would no longer be free to tell us what it means (that would be drawing the inference). Finally, we could require that scientists only step across the inductive gap when it is very, very small, and thus be as conservative in their risk-taking as possible. But why is this the right standard? Such a standard presumes that only risks of making a claim incorrectly matter, and ignores the risks of not making a claim when it is true, of waiting too long.

To attempt to be value-free in the assessment of evidential sufficiency is to ignore the broader society in which science functions, by being arbitrary, or ignoring the full set of risks, or ignoring the implications of scientific work in the broader society. If science is to be responsible to the broader society in which it functions, if it is to earn its authority, it should not be value-free at all. Instead, it needs to be value-responsive.

Suppose one still wanted to maintain the purity of science from social and ethical values, and that to do so one was willing to institutionally isolate scientists from society. This would involve not only making sure that only risks to scientists and to research were considered in the assessments of evidential sufficiency, but keeping scientists from saying anything publicly about their research. Others would need to maintain and police the border between science and society, deciding what bits of information, which pieces of scientific research, were ready for public consumption and which were not. Communication among scientists would need to fall behind a shroud of secrecy, insulating scientific meetings, publications, and debates from public consumption. Scientists could be free to pursue inquiry indefinitely, and someone else would need to decide when the evidence was enough to instigate other decisions or actions. Scientists would need to eschew the public eye, and would likely need to be physically isolated from the rest of society. We could sever science from society in this way, and thus keep scientists willfully ignorant of the societal implications of their research and from thinking about them. We could have others trained to do this for scientists and have those specialists deciding when evidence was sufficient for a public communication of a claim.

I think we should view such an approach with alarm, and indicative of a misplaced desperation to keep science "pure." Not only would such isolation likely produce questionable science (because the forums for discourse would have to be closed to only professional scientists, who would have to be more strictly credentialed than is currently the case, thus narrowing who was engaged in scientific discourse), but we would need to create and monitor an entirely new social institution. Who would keep track of the boundary policers, and whether they were acting in the public interest or corrupted by a narrower interest? These would be very difficult issues to address. It would also be a very authoritarian institution, as it would require the end of the free exchange of information, and sequestering of the entirety of empirical investigation under confidentiality wraps. The potential for abuse in such an institution is staggering. Despite the complexity we face with the demise of the value-free ideal, I think addressing the difficulties of relinquishing the value-free ideal is both more manageable and desirable than a truly isolated scientific enterprise.

Nevertheless, the demise of the value-free ideal does leave us with a problem in thinking about science and values: What ideals *should* guide the interaction of science and values?

Searching for New Ideals

That we need some ideals for values in science seems clear. Social and ethical values can have distorting and problematic effects on science, as evidenced by the cases of sexist science uncovered by feminists. Such cases are just one way in which social and ethical values can distort science. Occurrences of manufactured doubt show the influence of social ideologies on scientific research. Because the purveyors of doubt care so much about protecting unfettered capitalism, they are willing to distort the scientific record to forestall unwelcome policies.[15] Social values such as making a profit can lead scientists in the employ of for-profit entities to bend science (e.g., by selectively reporting the results of clinical trials in medical research).[16] And some cases of scientific fraud can be viewed as a pernicious influence of social values, when scientists are so sure of how the world should be, they make up the data to show that it is that way (e.g., the psychologist Cyril Burt and the manufacture of twin data to support his beliefs about the inheritability of intelligence).[17] Social and political values also drove such catastrophic cases as the influence of Trofim Lysenko on Soviet science under Stalin. We should not be sanguine about allowing social and ethical values into science unfettered. Such laissez-faire attitudes about values can make a mess of science.

Philosophers of science have offered several alternative ideals for thinking about how values should operate in science. I will articulate those ideals here and assess their strengths and weaknesses. We will see that there is no one all-encompassing ideal that can replace the traditional value-free ideal. What relinquishing the value-free ideal requires is that we grapple with a more complex terrain of science-society interactions.[18] Different ideals get at different aspects of scientific practice more or less effectively. Understanding their strengths and weaknesses allows us to see what they are useful for both philosophically and practically.

In the current literature (and I can make no claims to completeness in this fast-moving field), there are at least five different ideals (or norms) for values in science:

1. Placing priority on epistemic values
2. Role restrictions for values in science
3. Getting the right values into science
4. Ensuring proper community functioning
5. Ensuring good institutional structures for scientific practice

Let me describe each, articulating their strengths and weaknesses, and then we can see how they fit together.

1. Placing priority on epistemic values

Daniel Steel has suggested that the correct ideal for values in science is to make sure they do not hinder the attainment of truth (within the realm of "practically and ethically permissible" science).[19] Ethical values, of course, do restrict our methodologies and the kinds of science we pursue, so Steel does allow those kinds of restrictions on scientific research, even if they do hinder the discovery of new truths.

But aside from this restriction, Steel wants no social or ethical values to interfere with the attainment of truth.

This is an interesting ideal, but presents some problems for practical guidance in science. It can be hard in practice to know whether a particular value judgment (whether social, ethical, or cognitive) is helping or hindering the attainment of truth in the middle of a research project or scientific debate.[20] Part of the excitement of science is not knowing where the truth lies, so whether a value is helping or hindering can be quite unclear without the benefit of hindsight. In addition, one can wonder whether this is the right approach to take even in cases where social and ethical values do hinder the attainment of truth. What counts as ethically permissible science is an ongoing contested arena (as the debate over gain-of-function viral research shows).[21] Sometimes ethical values can inhibit the attainment of truth (because researchers are following their conscience) before the ethical debate is settled, and we might be quite happy about that in retrospect. In short, this ideal works well only when we have settled both what the truth is and what the ethical boundaries of permissibility are, which means guidance *in medias res* is lacking. And we might decide in hindsight that some truths are not worth having, given the ethical costs of getting them. This ideal seems primarily useful for retrospective examinations of scientific debates.

2. Role restrictions for values in science

In my work, I have emphasized distinct roles for values in science. I have argued that there are two roles for values in science: a direct role (where values serve as a reason to do something, and thus direct the decision) and an indirect role (where values serve to help assess whether the available evidence is sufficient for an inference or choice). I have argued that depending on where one is in the scientific process, different roles are acceptable. For example, a direct role for values is acceptable in deciding which research agenda to pursue (e.g., because the scientist cares about a particular issue) and in deciding which methodologies to employ (e.g., because a particular methodology is ethically preferable). An indirect role is acceptable in these instances as well. But at moments of data characterization and inference (the targeted terrain of the value-free ideal), I have argued that we can maintain scientific integrity while permitting social and ethical values by constraining such values to the indirect role only.[22] It is also an ideal that can help guide discourse on contentious scientific issues, as it allows for both the expression of values ("Because of this value, I find the evidence insufficient") and guidance for productive debate ("What evidence would be convincing for you?"). (I will return to this in Lecture 3.)

This ideal is a direct counter to the value-free ideal, and targeted as narrowly as the value-free ideal is on these "internal" inferential moments. As such, it has little to say about the direction of research agendas. Further, it cannot help much with methodology selection (or distortion). Finally, it is not much of an ideal in the sense of something to strive for. It is more of a minimum floor, which if one does not meet, one is doing really poor science (such as writing down the data one wishes were accurate or making inferences that one wishes were true). Although I think it is an important norm to hold, it will not suffice for guiding scientific practice.

3. *Getting the right values in science*

Several philosophers of science have argued in recent years that the important thing to focus on for values in science is making sure that the right values are influencing scientific research.[23] Such authors have taken an "aims-oriented" approach to the problem of values in science. Janet Kourany, for example, has argued for a "joint satisfaction" ideal for values in science—that only when a decision meets both epistemic and ethical criteria is it a good decision. Kevin Elliott has called attention to the multiple goals of science, including both epistemic aims and social aims.

There are several things to note about this approach. The first is that all the authors that champion this ideal take pains to express concerns for, and support of, the value of inquiry. Both the epistemic aim and the ethical/social aim must be met, for example, in Kourany's joint satisfaction ideal. So this ideal is not just about social and ethical values, but about valorizing the general purpose of inquiry and discovery as well. The pursuit of truth matters a great deal to those who argue for this approach.

The second is that this approach successfully addresses concerns about research agenda choices and methodological choices, about which the role-restriction norm has little to say. Because both roles for values are acceptable for these choices, that approach has no normative bite at these stages. Arguing about what the right values are is exactly on target for these choices. For example, in cases where the methodological choices seem to be made to guarantee preselected outcomes, the get-the-right-values-in-science ideal can say that the decisions improperly neglect the value of inquiry, and thus are improper decisions.[24]

Finally, the authors who support this approach tend to want the values utilized to be also the result of good inquiry—not necessarily of the same kind as empirical scientific research, but still informed by good empirical results and robust philosophical debate. Values are not mere contaminants in our process of inquiry with this ideal, but a strong support of it, as they too are open to inquiry.[25]

However, despite its importance, it is doubtful that this ideal is enough. First, what the right values are is often hotly and openly contested. How we know we have the right values can be unclear. So guidance for scientists in practice can be lacking. Second, at the moment of inference (the moment of central concern to the value-free ideal and to the role-restriction ideal), this ideal provides either inaccurate or incomplete guidance. What are we to do when evidence arises that seems to challenge our value commitments? Suppose (and I think this unlikely) that we discover men and women really do have divergent mathematical abilities. Do we reject the evidence because it does not meet the joint satisfaction of ethical and epistemic values? Suppose it is strong evidence (and so meets the epistemic criterion). Do we reject it because it does not fit with our ethical commitments? This seems to conflate the "is" and the "ought," and falls into the trap of wishful thinking and worrisome distortion that the value-free ideal was meant to ward off. It is also a case where the role-restriction ideal serves us well. We can say we want stronger evidence before we are willing to give up on our belief in the general equality of mathematical ability, and we can even say (one would hope) what such evidence should consist of. But rejecting the evidence because we do not like what it says is unacceptable. It is precisely this move that climate deniers often make, and we are rightly frustrated by that.

In short, for guiding scientists in practice, we need both of these ideals—the role-restriction ideal and the get-the right-values-in-science ideal—in operation, although at different levels of granularity. At particular moments of inference, getting roles right is important. And in general, having the right values is important. Indeed, one could justify the roles ideal in terms of the aims ideal—that valuing inquiry properly means, in part, keeping values in the right roles. But as noted above, there is often contention about what the right values are. To address this, we will need a broader communal perspective.

4. Ensuring proper community functioning

One of the weaknesses of the get-the-right-values-in-science ideal is that it is mute when we don't know what the right values are. What then? Or, what if the right values encompass a plurality of values, all legitimate, with good reasons to support them and reasonable disagreement among them? What kind of ideal can we articulate under these circumstances? Further, the previous ideals generally center on the impact of values on particular scientific choices. How can we ensure that the conditions that support the requisite critical debate and pluralistic reflection in science are in place?

Philosophers of science (led by feminists) have focused on describing the conditions for proper community functioning to address these concerns. Ensuring that one has a diverse scientific community—with clear forums for debate, expectations for the uptake of criticisms, and effective distribution of research efforts reflecting needed diversity so that alternative theories can be explored—serves to provide essential conditions for the robustness and reliability of science.[26] Such conditions also provide assurance that value judgments will be elucidated and examined within the scientific community, and that if there are disagreements about which values should be shaping research agendas, those debates can occur in an open and productive way. Having proper community functioning is essential to ensuring that, if there is general agreement on the values, the right values influence science, and, if there is not agreement on the values, some diversity of values will be deployed in making judgments in science.

Some minimum of effective community functioning is needed for producing acceptable science. But we can always do better along the ideals that philosophers like Miriam Solomon and Helen Longino provide for us. This set of ideals, focused as it is on how communities of scholars should work and distribute their efforts, complements ideals 2 and 3, which are more focused on how particular choices should be made in science. The communal functioning ideal calls for proper response and uptake of criticism, for example, but it is from ideal 2, from an articulation of how values can properly play roles in scientific reasoning, that we can see what proper response and uptake consists of. (It is not proper, for example, to say: "I don't accept that empirical claim because it disagrees with my values." It *is* proper to say: "I find that evidence insufficient because of my values and my concern over false positives, so I want stronger evidence before accepting that claim.") That we need ideals both for governing particular choices and for guiding communities should not be surprising. What none of these ideals address, however, is how the scientific community should interact with the broader (democratic) public.

5. Ensuring good institutional structures for scientific practice

While the social epistemological tendencies reflected in ideal 4 are useful for thinking about how we want our scientific communities to work, they do not help inform how the scientific community should think about its role and responsibilities to the broader society or how we want to structure the science-policy interfaces that so powerfully shape the pursuit and use of science. This area for ideals is the least developed.

It is on this kind of interaction that many ideals articulated by philosophers working on science policy have focused.[27] The trouble is that the science-policy interface is multifaceted, and philosophers have yet to grapple with all the facets in articulating an ideal. What constitutes good institutional structure is very much up for debate. I will address this in more detail in the next lecture.

For now, I hope I have shown that we need some set of nested ideals crafted from those described above. Ideal 2 is the most targeted response to the value-free ideal (both narrowly focused on inferences in science), but once we relinquish this ideal and confront the complexity of science in society, it seems obvious that no one ideal will suffice. Without the value-free ideal narrowing our focus, we have to think about and address all the ways in which values do influence science and consider how that should occur.

The Authority of Science and Ideals for Science

No one ideal for values in science will suffice. We need nested ideals, articulated for individual actors, communal practices, and science-society interfaces, in order to ground the authority of science.

The authority of science rests on the interlocking character of these norms. At the communal level, scientists are expected to continually question and critique each other's work. They are expected to respond to criticisms raised, and to hold no scientific claim above criticism. Such mutual critique is a minimum for granting science prima facie epistemic authority. The more diverse and reflective of the plurality of society the scientific community is, the more taken-for-granted assumptions and unexamined value commitments will (hopefully) be elucidated, the more authority science should have.

But community practices need good individual reasoning practices with which to operate. Maintaining the proper roles for values in science keeps values from acting in place of evidence, which will support the critical interactions needed in science. New evidence should always be able to contest old positions, and this can only happen if values are not used to protect desired positions from unwanted criticism.[28] A scientist can point to their values to argue for why they require more evidence to be convinced, but they can never point to their values to argue for why evidence is irrelevant to the claims they make or protect. Asking for more evidence drives the inquiry dialectic; holding claims above evidential critique does not.

Further, it is not just in individual reasoning integrity (right roles) and communal practices, but in some shared values (operating within proper roles) that science gains its authority in a democratic society. I will discuss this more in the next two lectures, but for now, note that getting the values right, particularly in the realm of policy-relevant science, strongly supports scientific authority. That scientists are investigating questions we care about, using methodologies that we find morally acceptable and targeted at what we are concerned with, and using values we share for assessing evidential sufficiency, can and should make a big difference for what we think is epistemically authoritative. Thus, elucidating the proper roles and proper values for science is part of what makes science authoritative, rather than undermining the authority of science.

Finally, the authority of science also rests on its raw instrumental success. Relying upon scientific understandings of disease (e.g., in the instance of communicable diseases) has greatly increased lifespans; relying upon scientific understandings of materials has greatly increased the range of what we can manufacture; relying upon scientific understandings of what we can transmit in the air has transformed communication; and so forth. It is this raw instrumental success that is probably at the root of most of the trust that society places in science. But as we will see in the next lecture, we are running into areas of science where success is not easily measured, especially in the short term, and the problems we are addressing seem more interrelated than ever. The challenge of science in democracy is still with us.

Implications

There is much work to be done in further fleshing out the ideals for individual, communal, and societal practices in science. We need these levels of norms to mesh together (at least somewhat), so that our communal expectations and societal practices do not place impossible burdens on individual scientists. We need to figure out how these norms align and how to encourage the pursuit of the ideals in real scientific practice.

But we also need to ensure that there is some space between what society might want and what scientists can pursue. While the full autonomy and isolation of science is undesirable, we also don't want a science that only tells us what we want to hear. Some space is crucial for the practice of science. Keeping social values out of a direct role at the moment of inference is part of maintaining this space. Allowing scientists to have a say about research agendas (and to pursue some research for curiosity's sake) is another.

Science cannot be just a mouthpiece for societal interests. If it becomes this, it will not have any claim to distinctive epistemic authority. While we need knowledge to help us pursue our social goals, we also sometimes need to know when such goals are not feasible or desirable (because of what else will come with their successful instantiation). Science needs to be able to tell us when we are running into such issues, to be able to "speak truth to power." This ability is central to its authority in practice. How to protect science's ability to do this, in democratic societies with their requirements of accountability, will be the topic of the next lecture.

Notes

[1] Whether it was an equal obsession for *scientists* I leave open. There is evidence that asserting the purity of science was central for at least some scientists. See, e.g., Heather Douglas, *Science, Policy, and the Value-Free Ideal* (Pittsburgh, PA: University of Pittsburgh Press, 2009), 64.

[2] Carl G. Hempel, "Science and Human Values," in *Aspects of Scientific Explanation and Other Essays in the Philosophy of Science* (New York, NY: The Free Press, 1965), 81.

[3] Alison Wylie, *Thinking from Things: Essays in the Philosophy of Archeology* (Berkeley, CA: University of California Press, 2002); Emily Martin, "The Egg and the Sperm: How Science has Constructed a Romance Based on Stereotypical Male—Female Roles," in *Feminism and Science*, Evelyn Fox Keller and Helen E. Longino, eds. (New York, NY: Oxford University Press, 1996), 103; Patricia L. R. Brennan, Christopher J. Clark, and Richards O. Prum, "Explosive Eversion and Functional Morphology of the Duck Penis Supports Sexual Conflict in Waterfowl Genitalia," *Proceedings of the Royal Society B* 277, no. 1686 (2010): 1309–1314. See also Donna Haraway, *Primate Visions: Gender, Race, and Nature in the World of Modern Science* (New York, NY: Routledge, 1989); Evelyn Fox Keller and Helen E. Longino (eds.), *Feminism and Science* (New York, NY: Oxford University Press, 1996).

[4] Helen Longino, *Science as Social Knowledge, Values and Objectivity in Scientific Inquiry* (Princeton, NJ: Princeton University Press, 1990); Anne Fausto-Sterling, *Myths of Gender: Biological Theories About Women and Men* (USA: Basic Books, 1985); Sandra Harding, *The Science Question in Feminism* (Ithaca, NY: Cornell University Press, 1986).

[5] Helen Longino, *Science as Social Knowledge, Values and Objectivity in Scientific Inquiry* (Princeton, NJ: Princeton University Press, 1990); Helen Longino, *The Fate of Knowledge* (Princeton, NJ: Princeton University Press, 2001); Lynn Hankinson Nelson, *Who Knows: From Quine to a Feminist Empiricism* (Philadelphia, PA: Temple University Press, 1990); Heidi E. Grasswick and Mark Owen Webb, "Feminist Epistemology as Social Epistemology," *Social Epistemology* 16 (2002): 185–196.

[6] Kristen Intemann, "Why Diversity Matters: Understanding and Applying the Diversity Component of the National Science Foundation's Broader Impacts Criterion," *Social Epistemology* 23, nos. 3-4 (2009): 249–266.

[7] Whether diversity is generally effective at ferreting out value presuppositions is unclear, and probably context dependent. It has certainly helped in some prominent cases examined in the literature, but there is also evidence that implicit biases still

pervade scientific practice. Our own tacit value commitments are often opaque even to ourselves. But it does seem that some difference of perspective is needed to make such commitments apparent, whether that comes from within or without, even if the presence of such a difference provides no guarantee of its effectiveness.

[8] Isaac Levi, "Must the Scientist Make Value Judgments?" *The Journal of Philosophy* 57, no. 11 (1960): 345-357.

[9] On simplicity, see Malcolm Forster and Elliott Sober, "How to Tell When Simpler, More Unified, or Less Ad Hoc Theories will Provide More Accurate Predictions," *The British Journal for the Philosophy of Science* 45, no. 1 (1994): 1–35. On explanation, see Wesley C. Salmon, *Four Decades of Scientific Explanation* (Pittsburgh, PA: University of Pittsburgh Press, 1989), and Heather Douglas, "Reintroducing Prediction to Explanation," *Philosophy of Science* 76, no. 4 (2009): 444–463, for initial orientations. On prediction versus accommodation, a review can be found in Heather Douglas and P. D. Magnus, "State of the Field: Why Novel Prediction Matters," *Studies in History and Philosophy of Science* 44 (2013): 580–589.

[10] See, e.g., Thomas Kuhn, "Objectivity, Value Judgment, and Theory Choice," in *The Essential Tension: Selected Studies in Scientific Tradition and Change* (Chicago, IL: University of Chicago Press, 1997), 320-339.

[11] Helen Longino, "Cognitive and Non-Cognitive Values in Science: Rethinking the Dichotomy," in *Feminism, Science, and the Philosophy of Science,* Lynn Hankinson Nelson and Jack Nelson, eds. (Dordrecht, Netherlands: Springer, 1996), 39; Heather Douglas, "The Value of Cognitive Values," *Philosophy of Science* 80, no. 5 (2013): 796–806.

[12] Heather Douglas, "Reintroducing Prediction to Explanation," *Philosophy of Science* 76, no. 4 (2009): 444–463.

[13] Heather Douglas, "The Value of Cognitive Values," *Philosophy of Science* 80, no. 5 (2013): 796–806.

[14] See, e.g., Heather E. Douglas, "Inductive Risk and Values in Science." *Philosophy of Science* 67, no. 4 (2000): 559–579.

[15] Naomi Oreskes and Erik Conway, *Merchants of Doubt: How a Handful of Scientists Obscured the Truth on Issues from Tobacco Smoke to Global Warming* (New York, NY: Bloomsbury Press, 2010).

[16] Thomas O. McGarity and Wendy E. Wagner, *Bending Science: How Special Interests Corrupt Public Health Research* (Cambridge, MA: Harvard University Press, 2008).

[17] Stephen Jay Gould, *The Mismeasure of Man* (New York, NY: W.W. Norton, 1981); Raymond E. Fancher, *The Intelligence Men: Makers of the IQ Controversy* (New York, NY: W.W. Norton, 1985).

[18] One might say that we are now free to grapple with that more complex terrain, as a purely epistemic approach is no longer remotely adequate.

[19] See, Daniel Steel, "Epistemic Values and the Argument from Inductive Risk," *Philosophy of Science* 77, no. 1 (2010): 14–34; Daniel Steel and Kyle Powys Whyte, "Environmental Justice, Values, and Scientific Expertise," *Kennedy Institute of Ethics Journal* 22, no. 2 (2012): 163–182.

[20] As noted by Kevin C. Elliott, "Douglas on Values: From Indirect Roles to Multiple Goals," *Studies in History and Philosophy of Science Part A* 44, no. 3 (2013): 375–383.

[21] See Michael J. Selgelid, "Gain-of-Function Research: Ethical Analysis," *Science and Engineering Ethics* 22, no. 4 (2016): 923–964.

[22] Heather Douglas, *Science, Policy, and the Value-Free Ideal* (Pittsburgh, PA: University of Pittsburgh Press, 2009).; Heather Douglas, "Scientific Integrity in a Politicized World," in *Logic, Methodology, and Philosophy of Science: Proceedings of the Fourteenth International Congress*, Peter Schroeder-Heister, Gerhard Heinzmann, Wilfrid Hodges, and Pierre Edouard Bour, eds. (London, UK: College Publications, 2014), 253.

[23] See, e.g., Janet A. Kourany, *Philosophy of Science After Feminism* (New York, NY: Oxford University Press, 2010); Kevin C. Elliott and Daniel J. McKaughan, "Non-Epistemic Values and the Multiple Goals of Science," *Philosophy of Science* 81, no. 1 (2014): 1–21; Daniel Hicks, "A New Direction for Science and Values," *Synthese* 191, no. 14 (2014): 3271–3295.

[24] See also Torsten Wilholt, "Bias and Values in Scientific Research," *Studies in History and Philosophy of Science Part A* 40, no. 1 (2009): 92–101.

[25] This is a pragmatist, Deweyian view on value judgments. See Elizabeth Anderson, "Uses of Value Judgments in Science: A General Argument, with Lessons from a Case Study of Feminist Research on Divorce," *Hypatia* 19, no. 1 (2004): 1–24; and Matthew J. Brown, "Values in Science Beyond Underdetermination and Inductive Risk." *Philosophy of Science* 80, no. 5 (2013): 829–839.

[26] See Helen Longino, *Science as Social Knowledge, Values and Objectivity in Scientific Inquiry* (Princeton, NJ: Princeton University Press, 1990); Helen Longino, *The Fate of Knowledge* (Princeton, NJ: Princeton University Press, 2001); Miriam Solomon, *Social Empiricism* (Cambridge, MA: MIT Press, 2001); Kristen Intemann, "Why Diversity Matters: Understanding and Applying the Diversity Component of the National Science Foundation's Broader Impacts Criterion," *Social Epistemology* 23, nos. 3–4 (2009): 249–266; among others.

[27] See, e.g., Philip Kitcher, *Science in a Democratic Society* (New York, NY: Prometheus Books, 2011); Justin B. Biddle, "Can Patents Prohibit Research? On the Social Epistemology of Patenting and Licensing in Science," *Studies in History and Philosophy of Science Part A* 45 (2014): 14–23; Heather Douglas, "Inserting the Public into Science," in *Democratization of Expertise? Exploring Novel Forms of Scientific Advice in Political Decision-Making*, Sabine Maasen and Peter Weingart, eds. (Dordrecht, Netherlands: Springer, 2005); Heather Douglas, "Reshaping Science: The Trouble with the Corporate Model in Canadian Government," *Bulletin of the Atomic Scientists* 71 (2015): 88–97;James Robert Brown, "The Community of Science," in *The Challenge of the Social and the Pressure of Practice: Science and Values Revisited*, Martin Carrier, Don Howard, and Janet Kourany, eds. (Pittsburgh, PA: University of Pittsburgh Press, 2008), 189.

[28] As Elizabeth Anderson has argued, values in science should not lead to dogmatism. See Elizabeth Anderson, "Uses of Value Judgments in Science: A General Argument, with Lessons from a Case Study of Feminist Research on Divorce," *Hypatia* 19, no. 1 (2004): 1–24.

Commentary on Lecture 1

TENSIONS AMONG IDEALS

Kristina Rolin

In "Science and Values: The Pervasive Entanglement," Heather Douglas advances a research program on values and science. The research program aims to develop a set of ideals that are nested in the sense that some ideals are addressed directly to individual scientists, some others to scientific communities with the aim of providing guidance for interactions among scientists within these communities, and yet others to stakeholders with the aim of providing guidance for interactions among scientists, policymakers, and lay people. In order to promote this research program, I argue that there are tensions among some of the ideals Douglas recommends for scientists and scientific communities. Striking an appropriate balance between ideals and requirements that pull in opposite directions is crucial for the success of the research program.

In Section 1, I present an overview of the ideals and normative principles Douglas identifies in the literature on values and science. In Section 2, I introduce the ideal of cognitive diversity, which is thought to be part of the proper functioning of scientific communities. In Section 3, I argue that there is a trade-off between the ideal of cognitive diversity and the requirement of shared standards. Further, there is a tension between the ideal of cognitive diversity and the ideal of "getting the right values into science."

1. Five Ideals

In the literature on values and science, Douglas identifies five ideals. The first three ideals are norms that individual scientists should follow in their scientific practices. The fourth and the fifth ideals are descriptions of epistemically ideal social arrangements. These descriptions involve norms that guide scientists in their interactions with other scientists or stakeholders of science. Besides norms, they involve principles for organizing scientific communities and institutions:

1. Placing priority on epistemic values
2. Role restrictions for values in science
3. Getting the right values into science
4. Ensuring proper community functioning
5. Ensuring good institutional structures for scientific practice

Douglas argues that there is no one all-encompassing ideal that can replace the value-free ideal; that is, the view that non-epistemic values have no legitimate role to play in the evaluation and justification of knowledge claims. In her view, we need a complex set of ideals that includes not only the first three items on the list, but also norms and organizational principles from the fourth and fifth items on the list.

The first ideal states that non-epistemic values should not hinder the attainment of truth within the realm of morally acceptable science. While Douglas does not object to this ideal, she thinks that it is not very informative as long as it does not specify what morally acceptable science is. The second ideal states that moral and social values are allowed to play an indirect role in deciding when evidence is strong enough, but they are not allowed to play a direct role. While Douglas emphasizes the importance of this ideal, she reminds us that it has a rather narrow domain of application. The ideal is meant to give guidance for evidential reasoning, and it has little to say about other moments in scientific inquiry. Therefore, it is better thought of as a minimum requirement for good scientific practice rather than as a full-service theory on values and science. The third ideal states that when moral and social values play legitimate roles in science, scientists need to ensure that they are the right values. According to Douglas, one virtue in this ideal is that it "successfully addresses concerns about research agenda choices and methodological choices, choices about which the role restriction norm has little to say."[1] However, the third ideal is inaccurate and incomplete insofar as it does not tell us how people are to decide what the right values are, or how scientists are to be informed about the right values.

Whereas the first three ideals state norms that individual scientists are accountable to conforming to in their scientific practice, the fourth and the fifth ideals are concerned with epistemically well-designed scientific communities and institutions. The fourth ideal (ensuring proper community functioning) involves not just a single norm, but a set of norms that scientific communities need to comply with in order to be successful in the pursuit of their epistemic goals. An example of such a set is Helen Longino's social value management ideal as found in her book *The Fate of Knowledge.* According to Longino, scientific communities should conform to the four norms of publicly recognized venues, uptake of criticism, shared standards, and tempered equality of intellectual authority.[2] Longino claims also that "A diversity of perspectives is necessary for vigorous and epistemically effective critical discourse."[3] Another example of a set of norms intended to be applicable to scientific communities is Miriam Solomon's social empiricism. Solomon recommends that science policymakers take steps to cultivate cognitive diversity and dissent in scientific communities.[4] By cognitive diversity, she means a diversity of theoretical approaches that have some empirical successes.[5]

The fifth ideal (ensuring good institutional structures for scientific practice) involves, among other things, a set of norms that govern interactions between scientific communities and the broader society. Douglas observes, I think rightly, that there is plenty of work to do to develop this set of norms and organizational principles. The work involves answering such questions as: What are the responsibilities of scientists when they function as experts in society? What are the responsibilities of policymakers and lay people when they rely on experts or use scientific knowledge in their decision-making? What kind of institutional structures are ideal for facilitating interactions between scientists, policymakers, and lay people in different arenas of public life?

Douglas argues that we need a set of ideals crafted from all of the five items on the list insofar as the ideals form a consistent whole. In order to contribute to this research program, I argue that there are tensions among some of the ideals. Insofar as an epistemically ideal scientific community is thought to be cognitively diverse,[6] the ideal is in tension not only with the requirement of shared standards,[7] but also with Douglas's third ideal demanding that values in science are the "right values." To better understand the tensions, I explain first why cognitive diversity is seen as an epistemic ideal.

2. What Is the Ideal of Cognitive Diversity?

In order to understand why cognitive diversity is of epistemic interest, it is necessary to introduce a distinction between cognitive and social diversity. A scientific community is cognitively diverse when its members have, for example, different research styles and skills, different perspectives on the subject matter of inquiry, or access to different bodies of empirical evidence. A scientific community is socially diverse when its members have different non-epistemic values, such as moral and political values, or different social locations, such as gender, ethnic identity, nationality, and race. It is a matter of empirical inquiry to understand how social diversity might be connected to epistemically beneficial cognitive diversity.

A number of philosophers argue that cognitive diversity is epistemically beneficial because it maintains a distribution of research efforts in scientific communities, gives rise to critical perspectives, and generates new research problems. Cognitive diversity is not claimed to be an epistemic virtue intrinsically. The claim is rather that, under some circumstances, it promotes the epistemic goals of science when these goals are understood to include significant truth[8] or empirical success.[9] In this section, I present a review of arguments defending the epistemic benefits of cognitive diversity.

2.1 Distribution of Research Efforts

In Philip Kitcher's article "The Division of Cognitive Labor," cognitive diversity is understood as a *diversity of theories or methods* addressing a common problem. Kitcher argues that cognitive diversity is epistemically beneficial in certain stages of inquiry, when it is not yet possible to tell which theory (or theories) will be true or most successful empirically, or which method (or methods) will lead to a breakthrough. When competing theories have different epistemic virtues or when different methods have complementary advantages, it is more reasonable to distribute resources among the theories or the methods than to allocate all available resources to one theory or method.

Kitcher argues that a distribution of research efforts can be epistemically desirable even in an instance where it would be rational for all community members to agree that one theory is superior to its rivals. Kitcher suggests that at least some community members should pursue a theory that is widely known to be inferior to the most promising theory. While the pursuit of such a theory is not rational from an individual point of view (given a traditional conception of rationality), it can be rational from a community point of view. It is in the interest of the community to maintain a competition among rival research programs.

Kitcher also argues that even in an instance where community members are united in their understanding of theoretical virtues, a distribution of research efforts may be an outcome of scientists' personal interest in credit. Instead of evaluating merely whether a theory is acceptable in light of available evidence and background information, a rational individual makes decisions strategically by anticipating other community members' behavior. If an inferior theory turns out to be true, great credit will be due to those scientists who have risked their careers for it.

Kitcher's arguments have been developed further by many philosophers. For example, in "Scientific Rationality and Human Reasoning," Solomon argues that the geological revolution between 1920s and 1960s is an example of scientific change where cognitive diversity played an epistemically positive role by creating a distribution of research efforts. Unlike Kitcher, Solomon does not believe that a distribution of research efforts will take place by "an invisible hand of reason."[10] She thinks that science policymakers and scientists who are in a position to make funding decisions are responsible for ensuring that scientific disagreements are not closed prematurely.[11]

2.2 *Social Value Management*

In Longino's contextual empiricism[12]—or critical contextual empiricism[13]—cognitive diversity is understood as a *diversity of perspectives* on the subject matter of inquiry. While cognitive diversity does not always go hand in hand with social diversity, Longino suggests that in many cases, a diversity of perspectives is an outcome of a diversity of social values in scientific communities. For example, when feminist scientists entered the field of human evolution, they introduced a novel perspective on the anatomical and behavioral development of human species. In the controversy over human evolution in the 1970s, they challenged the "man the hunter" narrative by developing the "woman the gatherer" narrative to offer an alternative interpretation of empirical evidence. Neither perspective was apparent in light of empirical evidence. Both perspectives were value-laden in the sense that they assumed the centrality of one sex's behavior in the evolution of the entire species.[14]

In contextual empiricism, cognitive diversity is thought to be epistemically beneficial not only because it generates a distribution of research efforts, but also because it generates critical exchanges in the community. Criticism can improve scientific knowledge in many ways. It can help scientists identify and correct false beliefs or biased accounts of the subject matter of inquiry. And even when criticism does not give scientists a reason to reject a view, it can be epistemically valuable by forcing them to provide better arguments for their view or to communicate their view more clearly and effectively. Criticism can help scientists avoid dogmatism.

Longino argues that a diversity of social values is epistemically beneficial because scientists are more likely to identify values that have influenced scientific research when the values in question are different from their own. As she explains, background assumptions may be value-laden in the sense that they lead scientists to highlight certain morally and socially significant aspects of a phenomenon over others, or they have morally and socially significant practical consequences, such as promoting one conception of human agency over another.[15]

In order to keep the influence of social values at bay, scientific communities need to be constrained by the four norms of publicly recognized venues, uptake of criticism, shared standards, and tempered equality of intellectual authority. This is needed to ensure objectivity.[16]

2.3 Diversity of Social Experiences

In feminist standpoint theory[17]—or standpoint empiricism[18]—cognitive diversity is understood as a *diversity of social experiences* that have a bearing on scientific research. When cognitive diversity is understood in this way, it is closely related to a diversity of social locations. Thus, standpoint empiricism has affinities with social epistemologies that emphasize the epistemic benefits of democracy.[19] In both approaches, a diversity of social locations is seen as an epistemic resource because information that is relevant for understanding complex social phenomena is distributed across the society depending, among other things, on individuals' social class, occupation, education, gender, race, and ethnic identity.

Like many other philosophers, standpoint empiricists believe that cognitive diversity is epistemically valuable when it leads to a distribution of research efforts, critical perspectives, or novel lines of inquiry. In addition, standpoint empiricists argue that a diversity of social experiences brings yet another benefit to scientific communities: marginal or unprivileged social locations are potentially a source of insight on the way relations of power work in the society as well as in the production of scientific knowledge.

Standpoint empiricists argue that a marginal or unprivileged social location in and by itself may not have epistemically interesting consequences unless it is developed into a standpoint. In their view, a standpoint is a collective rather than an individual achievement.[20] Insofar as there is an epistemic advantage associated with marginal or unprivileged social locations, a scientific/intellectual movement is needed to realize the advantage. Scientific/intellectual movements are epistemically productive when they enable scientists to generate evidence under conditions where relations of power tend to suppress or distort evidence, and they provide scientists with an epistemic community where they can receive fruitful criticism for research that may be ignored in the larger scientific community.[21]

In sum, epistemically beneficial cognitive diversity can come in many forms – a diversity of theories, methods, perspectives, and social experiences – and have many causes. Cognitive diversity is thought to be epistemically beneficial for at least four reasons. One reason is that it generates a distribution of research efforts. As no one is in a position to know in advance which lines of inquiry will be fruitful, scientific communities are better off by distributing their resources among several different and sometimes competing theories and methods. Another reason to value cognitive diversity is that it is a source of critical perspectives, which can improve scientific knowledge in many ways. Critical perspectives are needed especially in those cases where scientific research is value-laden. Yet another reason to value cognitive diversity is that it is a source of scientific creativity that can lead scientists to pursue new lines of inquiry, search for new evidence, propose new hypotheses and theories, and develop new methods of inquiry. Finally, cognitive diversity is especially epistemically fruitful in research projects that aim to produce evidence despite obstacles raised by association with power and social inequalities.

3. Tensions Among Ideals

Douglas, like many other philosophers, thinks that normative approaches to values and science should be concerned with proper functioning of scientific communities. While there is some disagreement over what proper functioning involves, most philosophers emphasize the importance of publicly recognized standards. Shared standards are needed to ensure that theories, hypotheses, methods, and observational practices can be criticized in a meaningful way.[22] Such standards are expected in order to ascertain what counts as an appropriate criticism that deserves uptake and what counts as a satisfying response to the criticism. What exactly the standards are depends, of course, on the specialty and the discipline we are concerned with. The standards are not beyond criticism, but at least some of them need to be widely accepted so that scientists can come to agree on appropriate criticism and response to criticism.

I argue that there is a trade-off between the ideal of cognitive diversity and the requirement of shared standards. While the ideal of cognitive diversity is meant to ensure that scientific communities benefit from a wide range of critical perspectives, the requirement of shared standards sets limits to the amount of cognitive diversity scientific communities can accommodate. The reason for this is that the requirement of shared standards excludes those critics who fail to follow the standards of the scientific community, or at least a sufficiently large number of the standards. To what extent the shared-standards criterion limits the scope of appropriate criticism depends on how the criterion is understood.[23] The challenge is to understand how many and which standards need to be shared for a scientific community to be able to function as a forum for meaningful criticism and response to criticism. Meeting this challenge involves striking a balance between cognitive diversity and shared standards.

Besides this trade-off, there is a tension between the ideal of cognitive diversity and the ideal of "getting the right values into science." The latter ideal states that when moral and social values play legitimate roles in science, we need to ensure that they are the right values. Insofar as epistemically beneficial cognitive diversity is generated by social diversity, it seems that almost any social diversity should be welcomed into scientific communities. This has led some philosophers to worry, I think rightly so, that the ideal of cognitive diversity will invite morally and politically problematic views into science, such as sexist and racist beliefs. Clearly, this was not the intention behind Kitcher's, Solomon's, and Longino's arguments. Nevertheless, the concern is that despite good intentions, the ideal of cognitive diversity may be abused by sexists and racists to demand resources to scientific research that is complicit in sexist and racist ideologies.[24]

I argue that the tension between the ideal of cognitive diversity and the ideal of "getting the right values into science" can be reduced by giving more specific content to the latter ideal. If the latter ideal includes the requirement for tempered equality of intellectual authority,[25] then it is in conflict with sexist and racist ideologies, which violate the view that all human beings deserve to be heard and treated respectfully. The requirement of tempered equality does not protect those speech acts that undermine the requirement itself.

In response to the concern that the ideal of cognitive diversity invites problematic values into science, Daniel Hicks introduces the good faith principle. According to this principle, it is not sufficient to require that scientists play by the rules of scientific communities; they need to do so in good faith.[26] Good faith participation in scientific communities requires that scientists do not reject the moral-political principles that underwrite and motivate the norms of epistemic communities. Such principles, he argues, include formal egalitarianism and liberal pluralism.[27] While the former states that all community members enjoy the same formal standing, the latter insists that there is room for reasonable disagreement.[28]

4. Conclusion

I have argued that any attempt to arrive at a synthesis of the five ideals will have to consider trade-offs and tensions between ideals and requirements that seem to be in conflict. More specifically, I have argued that there are two tensions among the ideals: one between the ideals of cognitive diversity and the requirement of shared standards, and the other between the ideals of cognitive diversity and "getting the right values into science." The research program envisioned by Douglas needs to strike a balance between these ideals.

Notes

[1] Heather Douglas, this volume, 26.

[2] Helen Longino, *The Fate of Knowledge* (Princeton, NJ: Princeton University Press, 2001), 129–134.

[3] Ibid., 131.

[4] Miriam Solomon, *Social Empiricism* (Cambridge, MA: MIT Press, 2001), 150.

[5] Ibid., 117.

[6] Helen Longino, *The Fate of Knowledge* (Princeton, NJ: Princeton University Press, 2001); Miriam Solomon, *Social Empiricism* (Cambridge, MA: MIT Press, 2001).

[7] Helen Longino, *The Fate of Knowledge* (Princeton, NJ: Princeton University Press, 2001), 130–131.

[8] Philip Kitcher, *The Advancement of Science: Science Without Legend, Objectivity Without Illusions* (New York, NY: Oxford University Press, 1993).

[9] Miriam Solomon, *Social Empiricism* (Cambridge, MA: MIT Press, 2001).

[10] Ibid., 95.

[11] Ibid., 150.

[12] As exemplified in Helen Longino, *Science as Social Knowledge, Values and Objectivity in Scientific Inquiry* (Princeton, NJ: Princeton University Press, 1990).

[13] As exemplified in Helen Longino, *The Fate of Knowledge* (Princeton, NJ: Princeton University Press, 2001).

[14] Helen Longino, *Science as Social Knowledge, Values and Objectivity in Scientific Inquiry* (Princeton, NJ: Princeton University Press, 1990), 106–108.

[15] Ibid., 216–218.

[16] Helen Longino, *The Fate of Knowledge* (Princeton, NJ: Princeton University Press, 2001).

[17] For example, see Sandra Harding, *Objectivity and Diversity: Another Logic of Scientific Research* (Chicago, IL: The University of Chicago Press, 2015).

[18] For example, see Kristin Intemann, "25 Years of Feminist Empiricism and Standpoint Theory: Where Are We Now?" *Hypatia* 25, no. 4 (2010): 778–796; Alison Wylie, "Why Standpoint Matters," in *Science and Other Cultures: Issues in Philosophies of Science and Technology*, Robert Figueroa and Sandra Harding, eds. (New York, NY: Routledge, 2003), 26.

[19] See Elizabeth Anderson, "The Epistemology of Democracy," *Episteme* 3, no. 1–2 (2006): 8–22.

[20] Sharon Crasnow, "Feminist Standpoint Theory," in *Philosophy of Social Science: A New Introduction*, Nancy Cartwright and Eleonora Montuschi, eds. (New York, NY: Oxford University Press, 2014), 145.

[21] Kristina Rolin, "Values, Standpoints, and Scientific/Intellectual Movements," *Studies in History and Philosophy of Science* 56 (2016): 11–19.

[22] Helen Longino, *The Fate of Knowledge* (Princeton, NJ: Princeton University Press, 2001), 130.

[23] Kristen Intemann and Inmaculada de Melo-Martín, "Are There Limits to Scientists' Obligations to Seek and Engage Dissenters?" *Synthese* 191 (2014): 2751–2765.

[24] Daniel Hicks, "Is Longino's Conception of Objectivity Feminist?" *Hypatia* 26, no. 2 (2011): 333–351.

[25] Helen Longino, *The Fate of Knowledge* (Princeton, NJ: Princeton University Press, 2001), 131.

[26] Daniel Hicks, "Is Longino's Conception of Objectivity Feminist?" *Hypatia* 26, no. 2 (2011): 333–351.

[27] Ibid., 342.

[28] Ibid.

Commentary on Lecture 1

THE DESCRIPTIVE, THE NORMATIVE, AND THE ENTANGLEMENT OF VALUES IN SCIENCE

Matthew J. Brown

Heather Douglas has helped to set the standard for twenty-first century discussions in philosophy of science on the topics of values in science and science in democracy. Douglas' work has been part of a movement to bring the question of values in science back to the center of the field and to focus especially on policy-relevant science. This first lecture, on the pervasive entanglement of science and values, includes an improved and definitive statement of the argument from inductive risk, which she is single-handedly responsible for rehabilitating and returning to the center of the debate. This statement makes clear the fundamental and absolutely pervasive nature of inductive risk and its import for our understanding of the role of values in science. The lecture also provides a survey of the current field of alternative approaches to ideals for the epistemic role of values in science that is comprehensive and generous, yet critical of each. (The conceptual/semantic role of values in science, i.e., the role of so-called thick normative concepts, is unfortunately omitted—a point I will return to.)

In these brief comments, I focus on providing an alternative perspective on some conceptual and rhetorical issues in Douglas' account, specifically dealing with the nature of values and the relation of the descriptive and the normative. This will lead me to somewhat different evaluations of two of the five new ideals for values in science that Douglas canvasses in the lecture.

Clarifying "Values"

What are we talking about when we speak of "values" in discussions of science and values? Lack of clarity on this point seems to me to be one of the biggest lacunae in the literature. Although there has been much work on the nature of so-called epistemic values, and we've worried at length about the distinction between epistemic and non-epistemic values, when it comes to understanding what values are in general, or what "non-epistemic" or "social" values are in particular, there is a serious gap. Few have thought systematically about the nature of values and how to draw on ethics to better inform the philosophy of science.[1]

Consider the "problem of wishful thinking" that plays such an important role in our contemporary discussions of values in science,[2] including this lecture.[3] The worry, briefly, is that if values enter into science in the wrong way, we will be led to conclusions based on what we wish to be the case, rather than what is the case. The framing of the problem itself presumes that values are something like wishes, desires, hopes, ideals, or visions of how the world ought to be. Is that what values are, in the relevant sense? Certainly that is one element of values, but it is not the only one. We use "values" to refer to desires, evaluations, judgments, commitments, identities, ideals, and assumptions, as well as social and institutional structures. We talk about acts, events, objects, and agents as having value (intrinsic or instrumental), and that seems to refer not to hopes and wishes, but to attributes of those things here and now (although we might hope and wish for more of some and less of others based on their value). While philosophical liberals think of values in terms of the way the world ought to be, philosophical conservatives think of values in terms of preserving what we currently have and enjoy.

One dimension of the concept of values that I have been particularly keen to emphasize, following the work of John Dewey,[4] is the distinction between *valuing* and *evaluation*, *prizing* and *appraisal*, between what we happen to like or dislike and the values certified by a process of *value judgment*. Our pre-reflective values may lack evidential support, as do many of our habits, biases, assumptions, and received beliefs. Our value judgments, in the honorific sense of "judgment," are the product of reflection and inquiry, and as such (according to Dewey), value judgments must have evidential support. Our well-established value judgments have been tested in practice under a variety of situations, and led to success—social, practical, emotional, and scientific.[5]

Consider the example of feminist values. I say, "I am a feminist." This is a statement of identity. How I act, including how I act as a scientific inquirer, is shaped by how those actions relate to my identity. It involves a reflective commitment to claims such as, "Women deserve to be treated fairly," and "All persons, regardless of sex and gender, are equal." A feminist ideal is, for example, the hope for a world where everyone gets equal pay for equal work irrespective of gender. My feminism is not a received view (far from it, given my upbringing), but is supported by personal and social value judgments. Feminist values have a good track record of guiding science into productive channels, whereas sexist values have the opposite sort of record.

This fact lends feminist values further support. What's more, the introduction of feminist structures into many fields has had a positive transformative effect on those fields.

That's just some of the complexity lurking behind the simple term "values." This complexity matters in understanding the relationship between the descriptive and the normative, and how we evaluate ideals for determining the legitimate and illegitimate uses of values in science.

Relating the Descriptive and the Normative

One takeaway from the work on the science and values literature is that the much-discussed dichotomy between descriptive and normative, fact and value, "is" and "ought," is overblown. Douglas shows us some key reasons why that is, but still, in my view, gives too much credit to the distinction.

Consider the bromide that one cannot derive an "ought" from an "is." While it is true, strictly speaking, this sense of "derive" is unreasonably narrow and restrictive. In the same degree, one also cannot derive a generalization from a finite set of particular cases, nor can one derive a theoretical claim—say, about the charge of the electron—from a set of observations—say, of floating oil droplets.[6] The failure doesn't tell us much of interest about the relation between the two.[7]

Consider another way Douglas puts the concern about the relation between descriptive and normative: "Making a whole set of descriptive claims (with nothing else) does not make an adequate argument for a normative claim," nor vice versa, which again, is true.[8] But also, if Douglas is correct, a whole set of descriptive claims (with nothing else) does not suffice to make an argument for a (nontrivial) descriptive claim, because one must evaluate inductive risks.

Nor, except maybe in the most abstract ethical theorizing, does a whole set of normative claims (with nothing else) suffice to make an adequate argument for a normative claim. So it's a limited point.

There are also claims that sit uncomfortably astride the descriptive/normative dichotomy, those that involve what Bernard Williams called "thick concepts."[9] Concepts like "gender," "equality," "poverty," and "well-being" all have an inextricable mix of descriptive and evaluative content, and the attempt to regiment concepts into purely descriptive and normative ones distorts what such concepts mean and impair our ability to articulate, operationalize, and reason with such concepts.[10] What's more, thick concepts (and the explanations and models that incorporate them) play a central role in many parts of psychological, social, and biological sciences. When reasoning about and making judgments using such concepts, value judgments are always necessary.

What Douglas seems most concerned about is not the distinction between normative and descriptive, valuable and factual, per se, but a particular aspect of that distinction, between the ideal and the actual. Here we can draw a pretty sharp distinction, at least given a specific context, and for much the same reason that Douglas focuses on. It *would* be a grave mistake to confuse the ideal for the actual. This would be the epitome of wishful thinking, and a misunderstanding of the role of ideals. And yet, worthy ideals are also not rootless fantasies. They have what Dewey calls an "active relation" to the actual, which distinguishes them as worthy ideals rather than mere fantasies—a realizability, a partial realization, a concreteness gained by successfully guiding action and social effort.[11]

As argued above, well-established value judgments have both evidential support and success guiding action under their belt. Some have success guiding scientific inquiry as well. Thus, contra Hempel, in the moderate holism that is the lot of scientific epistemology, when such judgments enter into scientific inquiry, value judgments can lend their own support to evidence and hypothesis.

On Role Restrictions

Douglas is right that we need to rely on several of the ideals in the literature to distinguish legitimate from illegitimate uses of values in science. What's more, something must be right about her approach to delimiting specific roles for values in science. But I think there are two kinds of problems with the direct/indirect role distinction, and the way Douglas uses the distinction to restrict the roles for values in science. First, I suggest that we need a better basis for the role restrictions, but that basis, properly understood, problematizes some of the claims that Douglas makes. Second, I think the distinction is too simple to cover the various roles that values of different types should play in different phases of scientific inquiry.

According to Douglas, evidence or values play a direct role in science when they act as reasons in themselves. Values play an indirect role, instead, when they "help assess whether the available evidence is sufficient for an inference or choice."[12] Values can and should play a direct role when it comes to external decisions about research agenda or methodology, but only evidence should play a direct role in internal, inferential, justificatory moments, while values should be relegated to an indirect role. This restriction is intuitively plausible. Ordinarily, values seem not to be the right sort of thing to justify empirical claims, whereas evidence is. On the other hand, values do seem to motivate or justify actions or practical decisions, such as what research we choose to spend our time and money on, or how we ought to treat research subjects.

Several detractors have found the direct/indirect distinction and restriction obscure.[13] I suggest that we need to find a more basic justification for the plausibility of the distinction. I see two sources of justification. One is the distinction between *reasons to act* (practical reasons) and *reasons for claims* (theoretical reasons). Reasons to act provide motivations to do something or make a certain decision; they also justify, in the ethical or practical sense, the action taken or the decision made. Reasons for claims, on the other hand, stand in logical relationships to those claims; they justify them in the epistemic sense. What the inductive risk argument shows, in my view, is that no amount of reasons for claims compels us to jump the inferential gap from reasons *for* a claim to reasons to *assert* or *infer* that claim. Assertions and inferences are actions, and as such require values to justify them.

When it comes to assertion, reasons to act becomes a tricky matter. The *norms of assertion* are the second source of what's right in the role-restriction ideal.[14] Inductive risk is a necessary consideration in assertion, because one has a moral obligation to consider the perlocutionary effects of one's assertions.[15] In terms of role restriction, it is central to the norms of assertion that one should not assert something you know or believe to be false (the sincerity norm). This norm is of course, defeasible; sometimes it is permissible or even obligatory to lie, and a lie does not cease to be an assertion.[16] But one does so at the cost of corrupting the practice of assertion and the social relations that normally depend on it.

The worst cases of values playing a direct role, for instance, making a claim that flies in the face of the evidence, amounts to a major violation of the norms of assertion, such as the sincerity norm, or the requirement that one have evidence or be able to defend one's entitlement to assert the claim.[17] That one might do so on the basis of other norms of assertion does not mean that such assertions are unproblematic. Other putative violations of the direct role restriction may be less problematic, however, such as when one refuses to assert a claim that one has sufficient evidence for because of its problematic consequences, as remaining silent never violates the norms of assertion. Nevertheless, it seems that the norms of assertion, and likewise the role restrictions, are defeasible, if only in extreme circumstances and at significant cost. One does not fail to assert when one lies, and one does not fail to do science when one makes an assertion because it fits with one's values; but in both cases, one does something *pro tanto* wrong. Of course, in such a case, one does prima facie bad science and might even fail to do science at all, for example by failing to present any evidence. But the direct role restriction, justified in this way, is not an indefeasible condition on science as such.

When the direct role for values is permissible, it is because the values are providing reasons to act. When the direct role for values is impermissible, it is because we are talking about reasons for claims, and values don't provide reasons for claims (at least, they don't ordinarily do so, or are not thought to do so). The indirect role, I think, is based in the limitations on other values created by the norms of assertion, especially the sincerity and evidential conditions, whose violation is rarely justified, and cannot be engaged in without damaging the practice of assertion (and so the practice of science). But the norms of assertion may not always line up with the way Douglas has laid out the distinction between direct and indirect roles.[18]

Finally, is it always true that values cannot provide reasons for claims, that is, provide evidence or epistemic support? In some cases, that seems right: that I value honesty does not mean that I or any other person tells the truth regularly. On the other hand, there is a long tradition of feminist philosophy of science arguing that, in certain cases, values do provide evidence for claims.[19] Consider, again, feminist values. For many, these values are the product of value judgments—the result of a process of inquiry justified by evidence. Feminist values are not simple or precise, but a broad network of value claims, many of which involve thick concepts. Action based on these values has improved many lives in a wide variety of contexts. They also have a strong track record of success in guiding inquiry fruitfully and helping uncover flaws and biases in past studies. Certainly, I should not simply infer from these value judgments to claims that "People do receive equal pay for equal work irrespective of their gender" or "Women today are accorded as much respect as men." This would be to confuse the ideal and the actual. But can I infer that "Men and women have equal natural talents at math, or near enough for practical purposes"? Can I justifiably assert, on the basis of these values, that "This study that shows divergent mathematical abilities in men and women is probably flawed"? I am inclined to say yes to both, based on the particular nature and status of these value judgments, just as I am inclined to dismiss until further research comes in news of physicists observing superluminal speeds or reports about "miracle" drugs. Good value judgments, like well-grounded theories or seasoned experience, carry epistemic weight.

The direct/indirect roles distinction and the role-restriction ideal assume overly simplistic distinctions between discovery and justification, external and internal aspects of science, as well as too simple a view of the landscape of values. We need a more complex account of the roles for values, given a richer theory of scientific practice than the context distinction provides, and the varying types and statuses of values that play a role in science. That said, I think the role restriction ideal as Douglas describes it remains our best account to date of the roles of values in science. For all its flaws (if they are, in fact, flaws), none of the distinction's detractors have provided anything like an adequate replacement. In terms of practical advice, it provides the best framework on the table.

On Getting Values Right

Again, I agree with Douglas that something like the ideal of getting the values right is a crucial supplement to an account of the proper roles of values in science. However, Douglas lumps together some views that need to be carefully distinguished.

First, we need to distinguish views (1) where the "right values" are a categorical absolute;[20] (2) where the "right values" are hypothetically given by the aims of inquirers;[21] and (3) where *value judgment* is a process of determining the right values for the situation (which is the view I favor). If what we want to do is use the right values, understood as absolutes, then the contestation of values creates some significant uncertainty for practitioners, in the same way that "truth" creates problems for the prioritization of epistemic values.[22] On the hypothetical view, we take the values of the inquirers as given, and the contestation is really an external argument about which kind of inquiry we should pursue. On the other hand, what contestation tells a value-judgment-oriented view is that we need to look for the value judgments that are warranted by inquiry and those that are not, or, failing that, that inquiry remains to be done. It is true that to provide adequate guidance, such a view needs to provide a robust account of value judgment or value inquiry, but there are already a variety of thinkers working in this direction.[23]

Let me say one final thing about what we should do when our evidence seems to clash with our value commitments. Certainly, the right-values ideal is not committed to "rejecting the evidence because we do not *like* what it says."[24] If we reject the evidence, it is because we are convinced it is wrong. Is it possible to justifiably believe that the evidence is wrong (or misinterpreted) based on values? In certain cases of the type I have already mentioned, yes.

But the values in question are not mere desires and preferences, nor are they ideals or "pious hopes." They are reflective, evidentially warranted judgments.

Notes

[1] Justin Biddle and Dan Hicks are two exceptions. See Justin B. Biddle, "State of the Field: Transient Underdetermination and Values in Science," *Studies in History and Philosophy of Science* 44, no. 1 (2013): 124–133; and Dan Hicks, "A New Direction for Science and Values," *Synthese* 191, no. 14 (2014): 3271–3295.

[2] For example, see Elizabeth Anderson, "Uses of Value Judgments in Science: A General Argument, with Lessons from a Case Study of Feminist Research on Divorce," *Hypatia* 19, no. 1 (2004): 1–24; Matthew J. Brown, "Values in Science Beyond Underdetermination and Inductive Risk." *Philosophy of Science* 80, no. 5 (2013): 829–839; Heather Douglas, *Science, Policy, and the Value-Free Ideal* (Pittsburgh, PA: University of Pittsburgh Press, 2009), 87.

[3] Heather Douglas, this volume, 27.

[4] John Dewey, *Theory of Valuation*, in *The Later Works of John Dewey, 1925–1953, Volume 13, 1938–1939:Experience and Education, Freedom and Culture, Theory of Valuation, and Essays*, Jo Ann Boydston, ed. (Carbondale, IL: Southern Illinois University Press, 1988/2008), 189.

[5] Cf. Elizabeth Anderson, "Uses of Value Judgments in Science: A General Argument, with Lessons from a Case Study of Feminist Research on Divorce," *Hypatia* 19, no. 1 (2004): 1–24; Sharyn Clough, *Beyond Epistemology: A Pragmatist Approach to Feminist Science Studies* (Lanham, MD: Rowman & Littlefield Publishers, 2003); Sharyn Clough, "Solomon's Empirical/Non-Empirical Distinction and the Proper Place of Values in Science," *Perspectives on Science* 16, no. 3 (2008): 265–279; Sharyn Clough, "Radical Interpretation, Feminism, and Science," in *Dialogues with Davidson*, Jeffrey Malpas, ed. (Cambridge, MA: MIT Press, 2011), 405; Maya J. Goldenberg, "How Can Feminist Theories of

Evidence Assist Clinical Reasoning and Decision-Making?" *Social Epistemology* 29, no. 1 (2015): 3–30.

[6] Cf. Elizabeth Anderson, "Uses of Value Judgments in Science: A General Argument, with Lessons from a Case Study of Feminist Research on Divorce," *Hypatia* 19, no. 1 (2004): 5.

[7] Douglas also regards the reverse, no "is" from "ought," as a standard presumption, but that seems to me to be just wrong. For one, plenty of ought claims have descriptive content: "Maxine ought to tie her shoes" fairly straightforwardly implies that "Maxine is wearing shoes." It is also a pretty standard presumption that *ought implies can,* and this small addition permits inferences from *ought to is*. So, if it is true that X *ought* to F, it follows that X *is* capable of F′. And assuming *ought implies can,* it follows that one can also at least derive an "ought not" from an "is not."

[8] Heather Douglas, this volume, 11.

[9] Bernard Williams, *Ethics and the Limits of Philosophy* (Cambridge, MA: Harvard University Press, 1985). Though Williams coined the term, he did not originate the idea. Williams himself claims to have picked it up from Philippa Foot and Iris Murdoch in the 1950s. See Hilary Putnam, *The Collapse of the Fact/Value Dichotomy and Other Essays* (Cambridge, MA: Harvard University Press, 2002), 159 n.19, and Pekka Väyrynen, "Thick Ethical Concepts," in *The Stanford Encyclopedia of Philosophy*, edited by Edward N. Zalta (Winter 2016), available online: http://plato.stanford.edu/archives/win2016/entries/thick-ethical-concepts/, for discussion of potential historical antecedents.

[10] Hilary Putnam, *The Collapse of the Fact/Value Dichotomy and Other Essays* (Cambridge, MA: Harvard University Press, 2002); John Dupré, "Fact and Value," in *Value-Free Science?: Ideals and Illusions*, Harold Kincaid, John Dupré, and Alison Wylie, eds. (New York, NY: Oxford University Press, 2007), 27.

[11] John Dewey, *The Quest for Certainty: A Study of the Relation of Knowledge and Action*, in *The Later Works of John Dewey, 1925–1953, Volume 4, 1929: The Quest for Certainty*, Jo Ann Boydston, ed. (Carbondale, IL: Southern Illinois University Press, 1984/2008), 1; John Dewey, *A Common Faith*, in *The Later Works of*

John Dewey, 1925–1953, Volume 9, 1933–1934: Essays, Reviews, Miscellany, and A Common Faith, Jo Ann Boydston, ed. (Carbondale, IL: Southern Illinois University Press, 1986), 1.

[12] Heather Douglas, this volume, 24.

[13] See, e.g., Kevin C. Elliott, "Direct and Indirect Roles for Values in Science," *Philosophy of Science* 78, no. 2 (2011): 303–324.

[14] Paul Franco has recently argued that we should shift our attention from the role of values in belief or acceptance to assertion, though in a sense, Douglas has always emphasized assertion or "making claims" as well. See Paul L. Franco, "Assertion, Nonepistemic Values, and Scientific Practice," *Philosophy of Science* 84, no. 1 (2017): 160–180; and Heather Douglas, *Science, Policy, and the Value-Free Ideal* (Pittsburgh, PA: University of Pittsburgh Press, 2009), 70.

[15] Paul L. Franco, "Assertion, Nonepistemic Values, and Scientific Practice," *Philosophy of Science* 84, no. 1 (2017): 176.

[16] Andreas Stokke, "Lying and Asserting," *The Journal of Philosophy* 110, no. 1 (2013): 33–60.

[17] John R. Searle, *Speech Acts: An Essay in the Philosophy of Language*, (London, UK: Cambridge University Press, 1969); Robert Brandom, "Asserting," *Noûs* 17, no. 4 (1983): 637–650; Paul L. Franco, "Assertion, Nonepistemic Values, and Scientific Practice," *Philosophy of Science* 84, no. 1 (2017): 160–180.

[18] On the other hand, when Douglas claims that an indirect role is permissible for typically "external" decisions, I find it puzzling, because such decisions require reasons to act, but not reasons for claims, nor do issues of assertion come into play.

[19] Lynn Hankinson Nelson, "A Question of Evidence," *Hypatia* 8, no. 2 (1993): 172–189; Elizabeth Anderson, "Uses of Value Judgments in Science: A General Argument, with Lessons from a Case Study of Feminist Research on Divorce," *Hypatia* 19, no. 1 (2004): 1–24; Sharyn Clough, *Beyond Epistemology: A Pragmatist Approach to Feminist Science Studies* (Lanham, MD: Rowman & Littlefield Publishers, 2003); Sharyn Clough, "Solomon's Empirical/Non-Empirical Distinction and the Proper Place of Values in Science," *Perspectives on Science* 16, no. 3 (2008): 265–279; Sharyn

Clough, "Radical Interpretation, Feminism, and Science," in *Dialogues with Davidson*, Jeffrey Malpas, ed. (Cambridge, MA: MIT Press, 2011), 405; Maya J. Goldenberg, "How Can Feminist Theories of Evidence Assist Clinical Reasoning and Decision-Making?" *Social Epistemology* 29, no. 1 (2015): 3–30.

[20] Janet A. Kourany, *Philosophy of Science After Feminism* (New York, NY: Oxford University Press, 2010).

[21] See, e.g., Kevin C. Elliott and Daniel J. McKaughan, "Non-Epistemic Values and the Multiple Goals of Science," *Philosophy of Science* 81, no. 1 (2014): 1–21.

[22] Heather Douglas, this volume, 25–27.

[23] See John Dewey, "Logical Conditions of a Scientific Treatment of Morality," in *The Middle Works of John Dewey, Volume 3, 1899–1924: Essays on the New Empiricism, 1903–1906, 1899–1924*, Jo Ann Boydston, ed. (Carbondale, IL: Southern Illinois University Press, 1977/2008), 3; John Dewey, "The Logic of Judgments of Practice," in *The Middle Works of John Dewey, Volume 8, 1899–1924: Essays and Miscellany in the 1915 Period and German Philosophy and Politics and Schools of Tomorrow*, Jo Ann Boydston, ed. (Carbondale, IL: Southern Illinois University Press, 1979), 14; John Dewey, *Theory of Valuation*, in *The Later Works of John Dewey, 1925–1953, Volume 13, 1938–1939:Experience and Education, Freedom and Culture, Theory of Valuation, and Essays*, Jo Ann Boydston, ed. (Carbondale, IL: Southern Illinois University Press, 1988/2008), 189; Steven Fesmire, *John Dewey and Moral Imagination: Pragmatism in Ethics* (Bloomington, IN: Indiana University Press, 2003); Elizabeth Anderson, "Uses of Value Judgments in Science: A General Argument, with Lessons from a Case Study of Feminist Research on Divorce," *Hypatia* 19, no. 1 (2004): 1–24; Mark Johnson, *Morality for Humans: Ethical Understanding from the Perspective of Cognitive Science* (Chicago, IL: University of Chicago Press, 2015); Anthony Weston, *Creative Problem-Solving in Ethics* (New York, NY: Oxford University Press, 2007); Anthony Weston, *A 21st Century Ethical Toolbox*, 3rd ed. (New York, NY: Oxford University Press, 2013); Anthony Weston, *How to Re-Imagine the World: A Pocket Guide for Practical Visionaries* (Gabriola Island, BC: New Society Publishers, 2013); and Matthew J. Brown, *Science and Moral Imagination* (Pittsburgh, PA: University of Pittsburgh Press, 2020).

[24] Heather Douglas, this volume, 27, emphasis mine.

Lecture 2

SCIENCE AND DEMOCRACY: SQUARING EXPERTISE WITH ACCOUNTABILITY

Heather Douglas

Introduction: The Challenge of Science in Democracy

Science is neither a value-free endeavor nor a value-free product. Social and ethical values interweave themselves throughout the practice of science, from the choice of the direction of research, to methodological choices, to inference decisions, to dissemination and application decisions. We can trace scientific practice and see how the values entangled themselves in science. (Indeed, this is something historians and philosophers of science do quite well). Through such investigations, we can find judgments we disagree with, or evidence of improperly used values. Doing so, we can conceptually untangle evidence and values, but only at the cost of taking apart the claims made. When we want to remake the claims, we make new decisions and conduct new experiments—perhaps with different values this time—thus producing new claims. So while we can trace the influence of values on science, we cannot squeeze them out. They are woven into the fabric of science.[1]

We need to depend on this fabric, on the knowledge that science produces, for decision-making in democratic societies. Having as accurate and robust empirical information as we can for policy decisions is central to good governance. To ignore the empirical information that science produces would be folly, as has been shown in examples from Lysenko in the former Soviet Union, to HIV controversies in South Africa, to the slowness of the world's response to climate change. The eradication of smallpox and the gradual reversal of decline of the ozone layer and its beginning recovery are illustrations of the success of science informing governance.[2]

Science is important for its aid in producing instrumental success, giving us new understandings and capacities to act in the world. In addition to instrumental success, a well-run scientific endeavor also produces knowledge that calls for our attention and has the potential to shift our communal priorities, particularly in the discovery of unexpected impacts of our societal decisions and technological choices, and in the undermining of empirical presuppositions of long-standing ideological commitments. We need science to provide ongoing critical engagement with the empirical underpinnings of our understanding of the world.

So the science we produce should not *just* be a reflection of the values that we hold, be they tacit or explicit. Even with values interwoven throughout scientific practice, the role of the empirical, and the evidence that may not fit with one's expectations, creates the potential for the unexpected and the disruptive. And the practice of critical engagement among scientists should foster this potential. Ironically, the very thing that creates robustness in science—diverse actors engaging critically with each other over empirical information—creates instability in the knowledge we use for governance. The lack of fixed and permanent points in science is central to its robustness, but means the knowledge on which we depend will be inherently unstable, always revisable (even if never actually revised).

In addition, since the professionalization and specialization of science in the nineteenth century, science has become carved up into distinct areas of expertise. No one person can master all of science anymore. Even professional scientists are only experts in some areas of investigation, and the knowledge explosion that characterizes the past century demands specific expertise for specific problems. Such expertise is often opaque to nonexperts: hard to understand and even more difficult to assess. Though the languages of specialization help experts talk efficiently to other experts, they impede understanding for the broader populace.

Yet experts bring with them a facility with judgments (in their area of expertise) that can be invaluable. Genuine experts in their expert terrain have a sense of what is important, and what is not. Experts know which confounders must be controlled for, and how it should be done. Experts know which paths have been tried before and failed, and which remain open and enticing. Experts possess an awareness of their own ignorance, of the limits of their knowledge, as it is on the edge of those limits that they work.

Specialized expertise has produced tremendous empirical success, but it also poses a tremendous challenge to democratic governance. We need experts to inform our governance decisions, but we also need those decisions to be democratically accountable. We need experts "to be on tap, but not on top."[3] What does this mean and what *should* it mean in practice? We need governance to be assessable by the public, because that is who will make the recall decisions for the government. And we need those in power to be able to access and use expertise in a transparent way, so the public can assess what those in power are doing, particularly in response to the available expertise. How is scientific expertise to be utilized by the government in democratic societies, given the specialization of expertise, where it seems that experts can only be assessed by other experts? And, given the ineliminability of values from expert judgments and knowledge, how should science and democracy interact? How should the values that shape science be part of the democratic accountability of science?

The crux of the challenge of science in democratic societies is thus created by these three background conditions: 1) the reality of value-laden scientific knowledge and expertise, 2) the importance of scientific expertise for good decision-making, and 3) the general difficulty of assessing expertise for the nonexpert. We need to think through our democratic institutions, particularly those on the science-policy interface, to structure them so that science retains its robustness but is also accountable to the democratic society in which it functions. In this lecture, I will address two central kinds of institutions, science advising (part of science for policy) and research funding (part of policy for science). There are many other locations where science and policy—or more broadly, science and democratic politics—meet, including (and here is just a partial list): citizen science, crowdfunding scientific research, moral restrictions on scientific research, privacy concerns (especially in our big-data era), dual-use/dangerous research (and related restrictions on communication and publication of research results), and the scientific underpinnings of the democratic process (which is often built on census data). Obviously I cannot address all of this here, but I hope to show how to think in general about science-policy interfaces in democratic societies, and by doing so, provide a start on how to address the range of interfaces we have.

Before proceeding with a closer examination of expertise, we should set aside simplistic dreams of scientism, of science *determining* all of our decisions for us. Science cannot tell us what we should do—it provides the best empirical claims available with which to make decisions, but it cannot provide the normative claims with which to proceed. The idea that science can tell us what to value all on its own has occasionally been quite popular but those who make such arguments usually have implicit, and often quite controversial, normative commitments they smuggle into their arguments. Sam Harris, for example, presumes that the aim of social decisions should be to make people happy (to be assessed by looking at brain states), a clearly utilitarian perspective.[4] Whether this might be the wrong thing to do when it violates rights to autonomy is something to which Harris seems oblivious. Neil deGrasse Tyson's recent calls for a country of "Rationalia" similarly overlooks the need for clear normative commitments and decisions regarding thorny issues of justice in any well-governed society.[5] Science is not enough for such decisions, not even as value laden as it is. We need clearheaded social and ethical values for such decisions, and it is these that are at the center of both ethical discourse and (ideally) political debate.

We want science to *inform* what we do, but science cannot *tell* us what to do. How do we keep expertise on tap, but not on top? How should we evaluate the expertise we need (when we cannot become specialists in everything ourselves), and how do we assess the values that we know pervade expert judgment? Finally, how do we ensure that the expertise we need is being properly pursued, with integrity and relevant focus? I will begin answering these questions by first describing the nature of expertise, before turning to questions of how to hold experts accountable, and what we should hold them accountable for, in democratic societies.

The Nature of Expertise

What is expertise? As noted above, the cultivation of expertise generates a facility with a particular context, so that judgments are well attuned to what is known to work, what is known not to work, and where the uncertainties and complexities lie. Particular forms of expertise can be described along a spectrum, from those whose success criteria are readily assessed by the non-expert, to those for whom functioning well is difficult for nonexperts to assess. Examples of readily assessable expertise include chess masters (do they win their games?), magicians (do they pull off the illusion?), and car mechanics (is the car fixed?). We depend upon these kinds of experts to get things done, and when they are done well, we can (usually) tell. As Aristotle pointed out, the partaker of a feast can assess the quality of the food as well as the chef who made it, even if the partaker would be quite unable to produce such food.[6] And as Dewey pointed out, the wearer of a shoe can tell readily whether it hurts her foot, and thus evaluate the quality of the shoemaker's craft.[7] Some kinds of experts wear their capacity on their metaphorical sleeve for all to see.

But as we move along the spectrum, it becomes more difficult for nonexperts to assess whether the expert is successfully making judgments within their field. We might be able to assess when a weather forecaster is making good projections (did it rain or not?), but climate modelers are concerned with much larger scales (both spatially and temporally) than we readily experience. Whether their expertise is reliable is not assessable in so direct a fashion. Similar concerns arise for some of the most controversial areas of expertise, such as epidemiology (concerned as it is with population level effects and always hedged with concern for confounders) or ecology (where complexity, issues of scale, and confounders are common). When an area of science grapples with complicated real-world processes that cannot be easily isolated (spatially or temporally), and where science that pinpoints causal explanations accurately might be swamped by confounders in actual practice, expertise is harder to assess. Nevertheless, expertise remains crucial in these more complex contexts.

There are, of course, specializations that fall in the middle along this spectrum. Our healthcare providers (dentists and doctors, for example) can have their efforts to help us thwarted by confounders in practice, and thus it can take longer to assess success in these cases. Still, we usually stop going to a dentist who cannot help a toothache, or only seems to make it worse. We stop going to doctors (or we should) whose treatments exacerbate our symptoms, particularly over the long term. Readily assessable success is not a digital affair (either applicable or not), and it might be more or less difficult to assess success of a particular expert.

Thus, we can spread expertise out along a spectrum of whether nonexperts can assess the expert in terms of raw success. We can debate whether some expertise (e.g., economists) should be held to standards of raw success or not. But what should be clear is that some kinds of expertise, or cases of the application of expertise, should not be damned as faux expertise merely because the expertise failed to successfully guide our action or to accurately predict events. For some expertise, failure in particular instances of practice is not due to a failure of expertise, but due to real-world complexity, the action of confounders or unexpected events, or to the failure of population level information to apply properly to particular people. Failure in application is not always the expert's fault.

So what should we ask of experts in cases where raw success is not the correct measure of successful expertise? How do we know whether an expert actually has expertise? One fruitful account is from Julia Annas' discussion of expertise regarding moral knowledge. She writes:

> *Expertise requires that the expert, unlike the mere muddler or the person with the unintellectual knack, be able to "give an account"* (logon didonai) *of what it is that she is an expert in. The expert, but not the dabbler, can explain why she is doing what she is doing; instead of being stuck with inarticulacy, or being reduced to saying that "it feels right this way," she can explain why this is, here and now, the appropriate thing to do in these circumstances.*[8]

Here is what we should demand from experts whose success cannot be readily measured: Experts should be able to give an account, an explanation, for why they think what they think. They should be able to say what went into their judgments, where they felt the important judgments were, and on what basis they made them. It is not just the presence of judgment that makes someone an expert, but their ability to say *why* they think what they think that makes them an expert, particularly when success is not easily displayed.

What should inform expert judgment? And what should be on display (rather than success) when an expert gives an account of their judgments? Clearly, the epistemic underpinnings of expertise should be a substantial part of the account given. Current explanations for the phenomena at issue, and how the expert understands the phenomena, should be presented. For example, if an expert judges that a class of pesticides is a potential cause for a bee die-off, the expert should explain why, given their knowledge, that is a plausible judgment (because of the known effects of the pesticide, including how it kills pests; the nature of deaths; the timing of the problem; etc.).

But also on display should be the uncertainties and crucial points of judgment (particularly where other experts might judge differently). Thus, for this example, the expert should also explain what other factors might be at play, and why they view them as more or less plausible causal agents. Finally, part of the presentation of expert judgment should include the social and ethical values that shape expert judgment, in the directing of expert attention, in what methods are available for expert investigation (for example, some desirable methods are not available for pursuing information about the pesticide because the value of bees prohibits enough hives to be available for large-scale controlled studies on bees, and doing controlled studies in the field is extremely difficult as bees go where they will), and in the assessment of evidential sufficiency (through inductive risk). Our expert should make clear why they view the evidence as sufficient for any claims made, both in terms of the strength of the evidence and in terms of the risks they are concerned about if they make an inaccurate claim. Both aspects depend upon their expertise, but the second depends as well on the values they used.

Experts often make such values clear in their statements already. They make statements about concerns for under- or overregulation when discussing their work, or about waiting too long or acting prematurely. This is part of expertise, the judgments that are essential to expertise, and why we rely upon those experts. Giving an account of one's judgment as an expert thus involves both reference to the explanatory structure that underlies the area of expertise and to the social and ethical values that frame the problem and assess the sufficiency of the available evidence. Just as there is no value-free science, there is no value-free expertise. What does this mean for the accountability of experts acting as science advisors in democracies?

Science Advising: Accountability Mechanisms in Practice

How should experts be held accountable, through democratic mechanisms, for the judgments that are necessary for expertise to function? Describing accountability requires a description of who experts are accountable *to* and what they are accountable *for*. I will argue here that there are two groups to which experts giving science advice should be accountable: 1) their home expert community for the accuracy of their claims, and 2) the citizenry for the value judgments embedded in their claims. However, for full accountability for the value judgments in expertise, the expert community is needed as well, because whether an expert is actually deploying value judgments properly can often only be fully assessed by other experts.

Accountability to Experts for Accuracy

The first arm of accountability is clearly needed. Expertise, especially scientific expertise, is developed within an epistemic community, and it is that community that is perfectly positioned to assess the reasonableness and accuracy of an expert's claims. If an expert claims something that is obviously erroneous, it is other experts who are best positioned to notice and to call the expert into account.

This is precisely what did *not* happen in the case of L'Aquila. It is well known that science advisors were tried and convicted for providing a negligent earthquake risk assessment. (Their convictions have since been overturned.) While there are many things that can be said about this episode, two failures stand out. First was that the seismological experts that were part of the risk commission failed to correct statements made on their behalf about earthquake risk prior to the large earthquake that killed 300 people. A public official claimed that that the small tremors were decreasing the risk of a major quake, with the supposed backing of the risk commission. In fact, it was known that the opposite was true for that region. Small tremors increase the risk of a major quake, as was well known by the experts in the room. While they made no such public claim, neither did they correct the public official who did. As a result of the public official's statements, many people changed their usual behavior (which was to abandon their homes and spend the night in the town square when such small tremors occurred). It was this change of behavior, and the 30 traceable deaths that resulted, for which the experts were ultimately tried.[9]

The second failure was of the broader scientific community in response to the charges leveled at the scientists. The international scientific community was outraged when the scientists were charged, and news stories and petitions circulated that the scientists were being tried for "failing to predict an earthquake."[10] This, as the judge in the case made very clear, was simply not true. No one expected the experts to accurately predict an earthquake. What was expected was an accurate risk assessment, and this is what the experts failed to provide (either in the meeting or in response to the public official's statements). So both the local experts failed to correct an inaccurate risk statement and the broader scientific community failed to hold the local experts to account for their failure. The L'Aquila case shows the difficulties of experts holding each other publicly accountable for failures of expertise. When threatened, communities of experts tend to circle the wagons rather than to critically examine each other.

The L'Aquila case also shows that inaccurate claims, saying things that are known to be misleading or false, are not just a failure with respect to the expert community from which the expert comes, but also a failure for those whom the expert is advising as well as the citizens who need good advice. In this case, getting it wrong caused people to change their behavior, resulting in their death or the death of family members. So getting things wrong, when it is known that what the expert saying is wrong, is a failure for all parties involved. But it is the expert community who has to spot such missteps and correct them. It would be too much to ask that nonexperts generally spot and correct expert missteps (although it may be something nonexperts can occasionally do). Specialization, and its resulting endemic epistemic dependencies, requires that experts hold each other accountable for saying things known to be wrong.

It is crucial that expert communities resist the urge to protect their own when a member of their community errs and/or makes false claims. They need to show that they will hold each other accountable for claims made both within the expert community and in more public forums. It is on such critical practices that our reliance on expertise is (in part) grounded. Expert communities that fail to do this, particularly when claims are made in public realms, weaken substantially the integrity and reliability of the expertise upon which the public depends.

Accountability to the Public for Value Judgments

In addition to the accountability to the expert community for general accuracy and competency, experts need to be held accountable for the value judgments that are embedded in their expertise. This is a more complex accountability arrangement than basic competency and accuracy. In the next section, I will discuss the value judgments that shape which research is done, i.e., what experts research and thus develop their expertise about. In this section, I will focus on the making of claims by experts and the value judgments that are needed to assess the (already available) evidence. These value judgments, regarding which claims the evidence is sufficient for, are required to support the argument from inductive risk, and are central to the process of expert advising, when the expert must decide which claims to make given the available evidence. Experts should make clear the values informing these decisions, both so that accountability mechanisms to the broader public can work and also so experts can assist with the accountability of their fellow experts.

Before explaining how advisors can be held accountable for value judgments in their work, we need a general understanding of how science advising works in practice in democratic societies.

Science advising is a complex process that is institutionalized in many different ways across different political cultures and science-policy contexts. For example, there are science advisors embedded within government working for elected officials, science advisory committees struck by government agencies, and independent advisory bodies that are on call for government needs. None of these science advisory mechanisms is in general superior to the others. Rather, what is more important for science advisory mechanisms is that 1) they fit with the existing democratic practices and institutions of the country,[11] and 2) there be a robust set of advisory mechanisms available. I have yet to learn of a science advisory system where only one institutional mechanism sufficed. Government is complex, and most contemporary governments will require science advisory mechanisms that reflect that complexity.

That said, we can characterize two general kinds of advising mechanisms -- informal vs. formal science advice -- both of which are needed. For formal science advice, a committee must be struck to address a particular question or provide comment on an issue or report. The committee is often interdisciplinary in nature (as most interesting questions these days are not adequately addressed within one discipline). The purpose of the committee is to provide a way for the experts to discuss an issue with each other, to explore the relevant literature, to draw on the diverse expertise represented in the group, and to draft a report reflecting what the committee thinks in response to the instigating question or issue. This is rarely a quick process, usually requiring a few months at least. Examples of advisory mechanisms in this category include national academy of science committee reports (whether from the US National Academy of Sciences, the Canadian Council of Academies, or the UK's Royal Society), standing advisory committee reports (such as the Environmental Protection Agency's Science Advisory Board in the US or the Intergovernmental Panel on Climate Change), and officially sanctioned ad hoc committees. Any committee struck that drafts an official report should be considered formal science advice. The reports are often authoritative (if done well), providing a comprehensive look at the relevant literature and helping policymakers and politicians make sense of what to think about the available evidence (and where future research might fill crucial gaps).

This kind of advice can be contrasted with informal advice—that which occurs when a science advisor speaks one-on-one with a democratically accountable individual (usually a directly accountable decision maker, such as an elected politician or a political appointee). The science advisor is frequently someone of the advisee's choosing, and the relationship of advisor to advisee is one built on personal trust (if it is working well). The advice is rarely made public, and this is crucial to its ability to assist government officials. Part of what makes informal advice so effective is the ability of the science advisor to correct ill-informed understandings of technical issues held by the advisee. The advisee needs to feel free to express potentially naïve, or outright dumb, ideas, and to hear what the advisor has to say about them. Chief science advisor positions are paradigmatic of this kind of structure for science advising.

These two kinds of science advice require different accountability mechanisms. For formal advice, the committee can be democratically accountable through both the representation of different perspectives among different members of the committee and through the expression of the value judgments needed to come to the conclusions of its final reports (even divergent value judgments). Members of the public can then assess the reliability of the expert advice by examining whether an appropriate range of expertise and perspective are embodied by the committee and whether the value judgments expressed by the committee in its reports are acceptable. Indeed, these two accountability mechanisms for formal advice can reinforce each other in practice, as a more diverse committee is more likely to explicate differences in value judgments among members and air those in their deliberations; thus, the final report is more likely (although not guaranteed) to be both explicit and accurate in describing those commitments, enabling both the advisees and the general public to see how expertise is operating.

For informal advice, the advice often remains hidden, publicly inaccessible. And as the advice comes from one person, it is unlikely that the values will be as well explicated. But here the traditional mechanisms of democratic accountability (through electoral politics) come to the fore. The advisee is often an elected official (or directly appointed by an elected official), and part of what we elect people for is for their values, their character, and the nature of their judgment. The advisor is thus held accountable to the democratic public through their close relationship with the advisee, who is held directly accountable to the democratic public.[12] The advisee should pick an advisor whose judgment she or he trusts, and this is a matter of both competency and some shared values. The advisor need not have identical values, but there usually needs to be some overlap, so that what is of concern to the advisee gets properly addressed by the advisor.

This accountability through shared values with the public official does not mean, however, that the advisor should just say things the advisee will agree with. An informal science advisor should not be a yes-man or -woman. Values should not dictate what the advisor thinks; the empirical evidence still matters a great deal. Having an advisor who is broadly respected by the expert community from which they are drawn will thus be crucial, as that will create some accountability to the expert community for the accurate evidential basis of the advice given. And it is for this kind of independence that some informal advising systems work well with appointments that last through multiple governments, i.e., multiple elected officials. But if the values of the elected official clash with that of the advisor, that is a good reason to find a different advisor—both the trust relationship and the accountability mechanisms for the advising need at least some shared values.

One might be wondering, given this difference in kind, where the L'Aquila risk commission is on this topography. That was clearly a science advising body, a committee in form, yet because it was not drafting a report, it provided something more akin to informal advice than formal advice. Which brings us to a third kind of advising mechanism: advice for crises. When time is short and problems potentially acute, a full formal science advising mechanism is often impractical. But getting advice from one person will be incomplete. In such cases, committees are often struck to provide advice in response to immediate, short-term concerns. For such committees, lengthy reporting and drafting processes get in the way of providing timely advice. For these kinds of committees, representation of different perspectives (both disciplinary and value-based) becomes even more important. (It does not seem the L'Aquila risk commission had this kind of diversity, which might also be why it failed in its advisory function.)

In short, we have some plausible accountability mechanisms for the value judgments embedded in expert advice across several advising structures. The important thing is that at least one of the mechanisms be operational when science advice is given. If there is no such mechanism, it is too easy for science advice to be abused and for the mantle of scientific authority to cloak ideological commitments or political pandering. If, for example, formal advice is not made public and yet still has the appearance of a thorough examination of the issue by a strong committee, a politician can claim any preferred policy is supported by the advice, whether or not that is true (as Nixon did regarding the Presidential Science Advisory Committee report on supersonic transport in 1969).[13] If, for example, informal advice is given privately, but no basis of trust exists between the advisor and the advisee, the advice is unlikely to be heeded, and again the advisee can claim the mantle of scientific expertise having informed the final decision. While no system is immune from potential corruption, thinking through and instantiating some accountability mechanisms at least reduces the ease of possible abuse. Having some accountability structures recognized and in place helps ensure that science advice is utilized properly in the democratic process.

In addition to ensuring accountability mechanisms for each advising structure, the potential for corruption and abuse is reduced by having a robust plurality of advising mechanisms. For example, having formal advice produced both within and without the government (from standing government committees and external bodies such as national academies), and having that advice be made public, allows both experts to keep an eye on each other's work for accuracy and the public to see whether different experts come to very different conclusions.

Such a plurality also assists with a third aspect of accountability, where experts are again needed to hold each other to account for judgments made, particularly in formal advice. This third aspect depends upon the transparency of value judgments in advising reports, because whether the judgments made regarding the evidence in fact reflect the stated value judgments of experts can often only be assessed by other experts. If, for example, an expert panel claimed it was very concerned about the impact of a chemical on human health, but only utilized studies that involved animal models generally known to be insensitive to the chemical, the public will not be able to tell that there is something seriously amiss. Only other experts can tell when particular selection criteria or modes of analysis do not in fact reflect choices guided by the stated values. So here too the expert community will need to hold each other accountable for the reasonableness of judgments made in the crafting of formal advice, both in terms of accuracy and in terms of whether the judgments reflect the stated concerns of the expert panel.

Democratic accountability for expertise thus makes substantial demands on experts. Experts need to: 1) hold each other accountable to the available evidence (so experts don't make false claims), 2) make value judgments explicit in their assessments of the available evidence, and 3) make sure that judgments made by their peers are actually reflective of the value commitments stated by their peers. It is essential that expert communities function well and are openly critical of each other in order for these accountability mechanisms to work. Doing so is also essential to the pursuit of good science—here is where the demands of science and democracy align. Nevertheless, the life of an expert doing policy-relevant work in a democracy is not an easy one.

Alternative Paths?

At this point, the reader might be thinking that this is all needlessly complex. Why can we not find a more straightforward way to assess experts and hold them accountable? For example, one might be tempted at this point to hold experts accountable based on success—does their advice help produce the desired outcomes or not? While this may be a useful approach in some cases, in many cases such straightforward success criteria are not appropriate. Not only might the best expert advice still fail to produce desired outcomes if carefully applied in practice due to confounding causal factors in the world (things often do not go according to plan), but we should also not hamstring our elected decision makers by requiring them to follow expert advice exactly. Such a requirement would eviscerate democratic decision-making, which should always allow leeway for rejecting (or only partially accepting) expert advice if countervailing factors are involved.

It is thus perfectly legitimate for politicians to argue (and act on the arguments) that individual freedoms are more important than public health benefits (as has been argued in the case of vaccines or fluoridation policies), or that inexpensive and/or nationally produced energy is more important than climate change. If the politician thinks that these are the correct value trade-offs, they should make that argument publicly, and then be voted in or out of office accordingly.

It is also perfectly legitimate for politicians to argue that in some cases, they find the available evidence insufficient for a claim because of their values. If, for example, despite strong evidence for anthropogenic climate change, a politician argued that he did not think it strong enough to override his concern for maintaining cheap and affordable energy, that would be a legitimate reason to reject expert advice on climate change. Note that such a position is much more honest than current claims that simply reject the strength of evidence on climate change. Rather than outright rejection, arguments centered on whether the evidence is strong enough focus on whether residual uncertainties are acceptable. It also leads to a more productive political debate, as the politician can be challenged 1) to say what evidence would be sufficient to change his mind, 2) on whether the status quo is in fact more affordable, and/or 3) to say why the current generation matters so much more than future generations (a key ethical value at stake). Rather than blocking political discussion by reducing avenues for productive engagement, legitimate use of values and evidence in democratic decision-making opens it up. (More on this in the next lecture.)

For now, it suffices to note that often we cannot use the raw success of expert advice as a metric for accountability for the expertise, both because of the complexity of the contexts in which experts are needed and because ultimate decision-making authority must rest with democratically accountable officials.

Another gambit is to directly elect our experts to advisory positions, thus holding the experts democratically accountable. I suspect this would fail to work in practice for two reasons. First, the need to campaign for office would take a lot of time away from experts, and with this burden it is doubtful they could maintain the much-needed ties to their home expert community, thus maintaining their expertise. Second, the expertise needed (i.e., the particular specializations required) changes on a case-by-case basis. We must be able to draw on different specializations as the need arises, and requiring that experts be elected to expertise-based decision-making offices would be a slow and cumbersome way to ensure democratic accountability for expertise.

Because these more direct measures for expert accountability are not available to us, we need the complex (and sometimes diffuse) accountability measures described above. Experts need to be willing to be critical of each other, for inaccurate claims or judgments. Experts also should make their value judgments clear as part of giving an account for why they think what they think. Doing so will help with both accountability to the public and accountability to their expert community.

But how do we ensure that we have the needed expertise on hand when advice is required? How do we shape what is studied so that the appropriate and required expertise is likely to be available when we need it? It is to issues of policy for science that I now turn.

Research Funding: Accountability in Knowledge Production

Specialization of expertise has allowed for the detailed and rich production of knowledge that is now embodied by many experts. Over the course of the twentieth century, the public purse became increasingly important for funding scientific research. And even with the worldwide rise of privately funded science again in the late twentieth and early twenty-first century, the public funds at least one-third of the research pursued, and sometimes much more than that.[14]

Private funders (particularly those in the for-profit sector, i.e., not NGOs and foundations) have the most capital to spend on research, but, unsurprisingly, they spend it in ways that primarily serve their interests. While some of this funding does serve public interests (for example, by producing desired consumer products and new jobs, thus driving desired economic growth), this research will be deliberately blind to public concerns in some important ways. In particular, this research tends to be geared toward products that are patentable or over which intellectual property rights can be asserted. In addition, this research tries not to discover diffuse harms of private enterprise. One cannot expect private companies to fund research into detecting subtle environmental harms or subtle health hazards of its products.[15] Therefore, privately funded science will be geared toward the private interests that fund it.

That there is a public interest that diverges from private interests can be seen in two general kinds of things that science can discover. The first is that science can discover important causal processes for improving our lives that are not readily patentable. As James R. Brown has pointed out, we lack systematic and careful research into the effects of diet and exercise on our health, especially when compared with pharmaceutical research.[16] And what nutritional research is carried out is usually funded by the food industry, looking to support narrow claims for improving specific product marketability. This is because it does nothing to serve private interests to know, for example, whether a particular (but generic) change in exercise or diet is a more effective treatment for a disease than a particular drug. It is difficult or impossible to assert patent rights over such things, and thus to increase your company's profits with such knowledge. Yet this is precisely what the public and our medical caregivers need to know.

The second kind of research is when science discovers a public impact of private actions, especially unexpected harmful impacts of ostensibly private choices. Such discoveries can radically alter what we think of as private vs. public. One obvious example is the discovery of climate change – the discovery that burning fossil fuels impacts the entire world's climate makes the burning of such fuels a matter of pervasive public concern and not just an issue of private interest. Earlier in the twentieth century, discoveries of the impact of air pollution on public health and the impact of sulfates and nitrates on ecosystems through acid rain similarly made previously private decisions about energy sources a matter of public concern. Another example was the discovery of the health effects of secondhand smoke, turning a private decision to smoke cigarettes into a public issue that led to restrictions on smokers in public spaces.

To see how this might work today with an issue that is currently not so politically charged, imagine that scientists discovered that Wi-Fi systems had negative health impacts (such as increasing glucose intolerance). What was largely a private affair (how strong your Wi-Fi signal is) would then become a public issue, since your Wi-Fi would negatively affect anyone within exposure range. Regulations about how strong Wi-Fi systems could be would likely result. Science in this way can expand the purview of the public, and thus the purview of government.

Science does not always expand the realm of the public—it can contract it as well, by discovering there is no public impact for choices once of public concern. This has happened with the adoption of children by gay parents, once an issue of profound public concern. But as studies failed to find any adverse impact on the children, it is increasingly a purely private matter. Thus, science need not always expand the realm of government intervention. But even when it contracts it, there is still palpable social change. I suspect that science's ability to intervene in our public discourse in this way is one of the reasons science is increasingly distrusted by social conservatives—science can shift our social map dramatically.[17]

What these two kinds of research show is that we can expect the public to need knowledge that private interests will neither fund nor pursue. It is thus imperative that public funding of scientific research attend to these public interest areas—the detection of causal forces on known public interests and the discovery of new public interests. However, the public purse cannot be wholly dedicated to such ends. Scientists also pursue research because they find a particular line of investigation inherently interesting, or because they think doing so will be theoretically revealing. Research for curiosity's sake must also be a part of the public funding of science (as it is rare that private funding supports such work). Sometimes the results of this type of research will be of profound interest to the public. The history of science is replete with examples of the serendipitous discovery of publicly important findings from curiosity-driven research.

How do we ensure that our research funding regimes address the public's need for these kinds of knowledge? First, it is clearly important to not just rely upon private sources for funding science and to not have public funding geared primarily toward private interests. This latter problem arose in the last years of the Harper government in Canada, when national grants to academic researchers were increasingly required to have privately secured matching funds and government laboratories were being shifted to pursue private interest (e.g., the National Research Council – a central government laboratory – was rebranded and reshaped into a "concierge service for industry").[18] Harper was dismantling support for both public-interest science and curiosity-driven research. Funding regimens must be robust for both public-interest and curiosity-driven science, disentangled from the enticements of private interests.

Still, researchers could primarily focus solely on what interests them, driven by internal disciplinary concerns, and ignore the public interest altogether. Indeed, it was a concern that much academic science was pursued along these lines that generated incentives for more private interest science in academia, such as the US Bayh-Dole Act of 1980. But policies like this did not incentivize public interest science. We want to allow scientists to pursue their research where their curiosity leads in general (that sort of freedom and the diversity of agendas it produces is essential to the scientific enterprise), but we don't want the scientific community to neglect research in the public interest. How should we manage this?

Currently, we use a range of mechanisms to encourage researchers to pursue public interest science. Funding agencies often have targeted areas that encourage capable researchers of shifting focus to specific areas of perceived need. Citizen groups (particularly patient groups) have prodded scientists to alter their research to address their concerns. Crowdfunding has become a viable way to pursue particular pieces of research, especially those that are likely to be of public interest—and this combined with interest groups can produce substantial funding (e.g., the ice-bucket challenge for ALS research). The general call for increased Responsible Innovation, where research agendas are crafted in consultation with the public in particular areas, can help to shape research trajectories. And collaborative research projects, where researchers work with public stakeholders over weeks or months to generate specific knowledge needed for public decision-making, can be very powerful in producing relevant and trusted knowledge.

The difficulty is that there is no one person, or even one group, that is responsible for (much less accountable for) the direction of research. Yet this is probably a good thing, because we don't want science to be so tightly planned that there are clear lines of accountability. Currently we have a mishmash of different avenues to try to encourage a diverse and publicly interested research agenda. One wonders, though, if we could do better in promoting public-interest science, instead of curiosity-driven or private-interest science. Perhaps an oversight committee, drawn from different disciplines but geared towards the public interest, could review the range of research being funded and pursued and see if there are gaps that need to be filled. Perhaps such a committee could take comments and recommendations from the general public on whether there are areas where the public needs scientific expertise but is currently not getting it. Perhaps the committee could make recommendations to granting councils and agencies on where efforts are needed, and perhaps scientists could see if they could fulfill such needs. Such a system would not plan all of scientific research (even all publicly funded research), but it could nudge it in certain directions when needed. While it might be a good idea to pursue such a mechanism, it would be a mistake to allow it to replace the other mechanisms already in existence. This is partly because there will always be gaps and failures, and if we rely upon just one institutional structure, we will think that that institution is taking care of the concern so we don't have to worry about it anymore. Additionally, single institutions are always open to capture by special interests. Maintaining a pluralist system for public research funding is likely to be both desirable and necessary going forward.

One final point: What is clearly needed among scientists is more attention to not just the public in general but the least well off members of the public. The pervasive neglect of environmental contamination, health problems, and food insecurity among the most vulnerable in our communities is a deep problem, and a source of inequality that is a threat to both democracy and science. Public trust in science is shaken by the failure of academic scientists to pay attention to environmental and health problems, particularly when those problems turn into a crisis (e.g., water quality in Flint Michigan). Issues of distributional justice need to be more of a central concern, particularly as we strengthen efforts in public-interest science.[19]

Conclusion

We cannot design institutions at the science policy interface that will be forever immune from corruption and abuse. Democracy is an ongoing experiment in governance, and powerful interests will always try to find new ways to game the current system in their favor. It is part of the price of being a citizen in a democratic system that one must always examine whether the existing structures are helping or hindering the discovery, expression, and pursuit of the public interest.

That this is an ongoing struggle, and a process we are never done with, does not mean we cannot say something useful now at a general level. Acknowledging the value-ladenness of expertise, its importance for governance, and its general inaccessibility in its full details articulates clearly the challenge of science and democracy, but also clarifies how to proceed. Lines of accountability for expert advice run through both expert communities (which should be diverse and interactively critical) and the general public (which should be able to assess at least the values that are part of expert advice). That there are these two directions for accountability for expert advice should keep us from sliding into technocracy. We do not have to simply "trust the expert." We can demand to know why they think what they think, see what their fellow experts think, and assess whether the values expressed are those we agree with. Further, we can demand that expert communities pay attention to the needs of the public (especially the least well off). But such demands are not equivalent to having experts give us (or our elected officials) the answers they would prefer. If experts fail to "speak truth to power" (as happens in egregious failures of expertise), experts can expect moral outrage from the public. And rightly so.

Being an expert thus requires a certain moral courage, to pursue knowledge that is not just potentially unpopular, but also needed. The public interest is often revealed and coalesced around new pieces of research. We need science to perform this function.

This lecture, however, has focused mainly on the experts and their relationships with those in power (advisees and granting agencies) and with other experts. The public has played scant role thus far. It is in the area of science communication that scientists and the public more fully engage each other, and that is the topic for the next lecture.

Notes

[1] As noted in Kevin Elliott's book, *A Tapestry of Values* (New York, NY: Oxford University Press, 2017).

[2] For success stories, see Susan Solomon, Diane J. Ivy, Doug Kinnison, Michael J. Mills, Ryan R. Neely III, and Anja Schmidt, "Emergence of Healing in the Antarctic Ozone Layer," *Science* 353, no. 6296 (2016): 269–274; Richard A. Kerr, "First Detection of Ozone Hole Recovery Claimed," *Science* 332, no. 6026 (2011): 160; and DeWitt Stetten Jr., "Eradication," *Science* 210, no. 4475 (1980): 1203.

[3] This expression about the place of scientific expertise dates as far back as at least the 1950s. See, e.g., James Killian quoted in the *Bulletin of the Atomic Scientists* (Feb. 1957), or Eugene Rabinowitch in 1959 in the same journal.

[4] Sam Harris, *The Moral Landscape: How Science Can Determine Human Values*, (New York, NY: Free Press, 2010). See also his TED talk from that year: https://www.ted.com/talks/sam_harris_science_can_answer_moral_questions.

[5] Marko Kovic, "Neil deGrasse Tyson's 'Rationalia' works as a thought experiment, but it fails as a manifesto," *Medium* (Aug. 14, 2016), available online: https://medium.com/@marko_kovic/neil-degrasse-tysons-rationalia-works-as-a-thought-experiment-but-it-fails-as-a-manifesto-2e8cad2cf35e.

[6] Aristotle, *Politics* Book III, Chap. 11. For an English translation, see Jonathan Barnes (ed.), *The Complete Works of Aristotle, Vol. 2* (Princeton, NJ: Princeton University Press, 1984).

[7] John Dewey, *The Public and Its Problems: An Essay in Political Inquiry* (Athens, OH: Swallow Press, 1927), Chap. 6.

[8] Julia Annas, "Moral Knowledge as Practical Knowledge," *Social Philosophy and Policy* 18, no. 2 (2001): 244.

[9] Stephen S. Hall, "Scientists on Trial: At Fault?" *Nature* 477 (2011): 264–269.

[10] Ibid.; Michael Yeo, "Fault Lines at the Interface of Science and Policy: Interpretative Responses to the Trial of Scientists in L'Aquila," *Earth-Science Reviews* 139 (2014): 406–419.

[11] Science policy systems can be quite different even among similar types of governments, e.g., across parliamentary systems. Even more pronounced differences arise when one considers democratic governments of different forms, e.g., parliamentary

government (as in the United Kingdom) versus divided government (as in the United States).

[12] If the advisee is not an elected official, but rather appointed by an elected official, this mechanism gets watered down, but still operates to some degree.

[13] Heather Douglas, *Science, Policy, and the Value-Free Ideal* (Pittsburgh, PA: University of Pittsburgh Press, 2009), 40.

[14] See https://stats.oecd.org/Index.aspx.

[15] Unless, of course, such research is required by regulation, and then it can be gamed in the company's favor. See Thomas O. McGarity and Wendy E. Wagner, *Bending Science: How Special Interests Corrupt Public Health Research* (Cambridge, MA: Harvard University Press, 2008).

[16] James Robert Brown, "The Community of Science," in *The Challenge of the Social and the Pressure of Practice: Science and Values Revisited*, Martin Carrier, Don Howard, and Janet Kourany, eds. (Pittsburgh, PA: University of Pittsburgh Press, 2008), 189.

[17] Heather Douglas, "Science and the Public Square: Why Conservatism Doesn't Get Along Well With Science," Science Progress.org (2013), available online: https://sepos.cal.msu.edu/wp-content/uploads/sites/18/2020/08/Science-and-the-Public-Squre.pdf.

[18] Heather Douglas, "Reshaping Science: The Trouble with the Corporate Model in Canadian Government," *Bulletin of the Atomic Scientists* 71 (2015): 88–97.

[19] This point has been intentionally made regarding the distribution of medical funding and global disease burden. (See, e.g., James H. Flory and Philip Kitcher, "Global Health and the Scientific Research Agenda," *Philosophy & Public Affairs* 32, no. 1 (2004): 36–65.) But it applies across other areas of research as well.

Commentary on Lecture 2

VALUES AND ACCOUNTABILITY IN SCIENCE ADVICE: THE CASE OF THE IPCC

Arthur C. Petersen

In this commentary essay, I would like to delve more deeply into an important case that illustrates very well the concerns raised by Heather Douglas in her lecture on the accountability of expertise. The case I am referring to is an example of the way the Intergovernmental Panel on Climate Change (IPCC) deals with scientific and political values and accountability.[1] I will connect to several of the points that Heather raised in her lecture, especially the question: How assessable is expertise?

I will focus on the example of expertise on the causes of climate change ("attribution"). I will make the argument that if you want to assess expertise, you will have to engage with an "extended" peer community.[2] Reflection on assumptions should lead experts to give an account of the epistemic underpinnings of their expertise. I will argue that IPCC reports do not do that enough.[3] In pushing scientists to give such accounts, one must realize that experts often do not like to hear that; in this sense, their expertise should be considered to be on tap, but they are not on top in terms of being free to decide how transparent they will be.

Figure 1. Climate-Change Attribution Figure in the Summary for Policymakers of the Third Assessment Report of the IPCC[4]

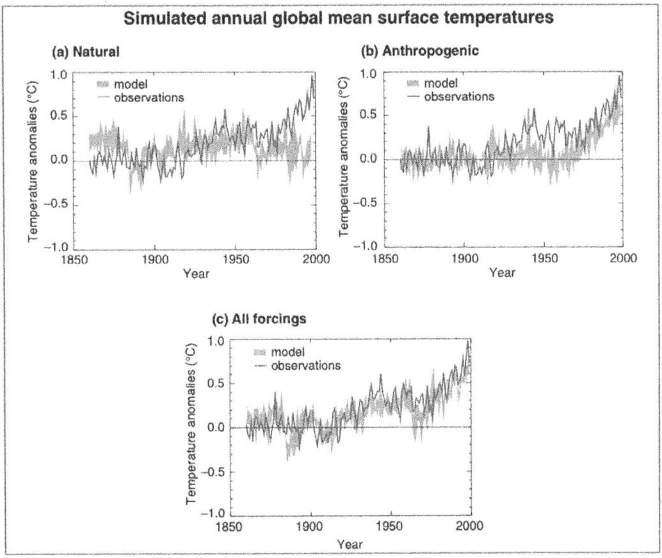

In the 2001 report of the IPCC, a figure was included that has become iconic at the science-policy interface for attributing climate change to human influences (reproduced here as Figure 1). The figure contains three panels, each showing, on the one hand, the same line with measurements of the global mean surface temperature since 1850 (going up in the beginning of the twentieth century and going up at the end of the twentieth century) and, on the other hand, a different band of model results (the bands representing the "internal" variability of the climate system, that is, the sensitivity to initial conditions): one for only natural external influences on the climate (volcanoes, sun), one for only human external influences on the climate (greenhouse gases, particles), and one that combines natural and human factors. The latter panel depicts a beautiful match of measurement and model, giving rise to the suggestion that we know everything, that there is no room left for any doubt that humans are causing the recent climate change. In fact, the chair of the IPCC suggested exactly this at a press conference in 2001.[5]

Of course, philosophers of science understand that the number of degrees of freedom in climate models is high. And they will not be surprised to hear that, indeed, virtually all climate-modeling groups in the world are able to present the same final panel with a match. This is not to say that the results are wrong. But how should one communicate that the bands are "just" model results, whose match with the measurements cannot establish reliability? The pertinent questions are: How do we know how reliable the models are? And in which senses can we say that they are reliable?

The IPCC has developed a methodology, through three subsequent guidances, for assessing and communicating the uncertainties in the findings of its assessments. This methodology includes calibrated terms for communicating probabilities. For the example of climate-change attribution to human influences, the IPCC did not communicate in 2001 that it was 100% certain that humans are causing climate change, even though the picture is beautiful and the line and band match. It said, rather, that it was "likely" that most of the warming of the last 50 years has been caused by human greenhouse gases. According to the experts, "likely" here means a 66% chance that the finding is true.

I was sitting at the table at the time (in Shanghai, on 20 January 2001) as an IPCC contact group negotiated what I think became one of the most important statements ever from the IPCC, that most of the warming is likely due to human influences.[6] But I could not understand why they said "likely." If you believed the models, the likelihood was already estimated to be way higher than 90% (that is, "very likely," the next likelihood category). I had to dig deep (through interviews, reviewing internal emails, etc.) to determine how the lead authors had reached their judgment. The reason they did not choose "very likely" was that they did not trust the models enough. So they picked the next lower likelihood category. Nowhere could this reasoning be found in the IPCC report; there was no traceable account of how they had arrived at this crucial judgment.

Six years later, the IPCC panel assessed the same question. The 2007 report features a similar figure as the 2001 report, but now the graphs are shown for every continent and the authors are willing to say "very likely" (90%). And again I could ask the question: Why not the next likelihood category of 99% or "virtually certain"?[7] The narrative could have been, "Even though we still do not fully trust the models, there have been more warm years, there have been more model runs, there have been different types of model experiments, and there is a belief that the models have become more reliable." I do think that the latter belief is problematic. Again the IPCC featured, in my view, a weak practice of assessing the reliability and the quality of models.

So what I argue has been missing from the Third and Fourth Assessment Reports of the IPCC (2001 and 2007, respectively) is sufficient attention to "methodological reliability" rather than simply "statistical reliability."[8] Assessment of methodological reliability requires a qualitative discussion and a corresponding qualitative assessment of the underpinning of results. Additionally, after "Climategate," the realization has come that "public reliability" needs attention too; how to gain back trust and be publicly relied upon is a difficult question for climate scientists. I do not have simple answers here. In this essay I am really focused on the importance of the second type of reliability: methodological reliability.

Let me give one example from the negotiations on representing methodological reliability in the *Summary for Policymakers* that occurred in Paris in 2007. This is the sentence that was under negotiation:

> *Most of the observed increase in globally averaged temperatures since the mid-20th century is very likely due to the observed increase in anthropogenic greenhouse gas concentrations.*

We have to get a bit into the politics now. Because these IPCC sentences are transferred from the sphere of knowledge assessment to the sphere of political negotiations (in the climate framework convention), there is always a country that does not want a stronger statement than the last time. A stronger statement would highlight that there is more scientific certainty, which would increase the likelihood of international agreements to curb climate change. The IPCC meeting in Paris in January 2007 was less than two years from what turned out to be the failure of the Copenhagen Summit at the end of 2009. In this instance, a country used all kinds of ways to prevent this sentence from being included. There is, however, an order of speech within the IPCC, which is: the chapters have been written – hands off, governments cannot touch those chapters! – but government delegates can comment, making use of a set of criteria (such as clarity and representativeness), on sentences in the Summary for Policymakers. Governments obviously will have different views. And the authors have a veto right on any change that is made to their summaries. One can imagine how hard it sometimes becomes to negotiate the summary line by line, as is the case in the IPCC. But it works.

Still, I argue that it can be done in a more productive way if both parties, authors and governments, would behave more diplomatically toward each other, understand better where they are both coming from, and what their respective rationalities are. One group of actors in these meetings is there on authority of their social, ethical, political, and economic values (their role is to represent their publics), and another group of actors is there on authority of their scientific values (their role is to represent, to the best of their ability, the papers they have assessed), and must provide good "reference."[9]

Back now to the sentence that was under discussion in the final hours of the Paris meeting. After days of negotiations and having entered very deep into the night, finally we are in agreement—all the countries of the world can agree on the sentence by inserting the following footnote: "Consideration of remaining uncertainty is based on current methodologies." Of course we were all tired. But it is interesting: Why would the opposing country agree with this sentence? What is the spin they could give? They might say, "The methodologies used are based on models. It is just models. It is not reality." Indeed models are used, but that does not imply that there is no reference to reality; still, that is typical of the argument they would make. How would another country that tends to dramatize climate change and typically wants to downplay uncertainty spin this sentence? They might say, "Next time the likelihood will go up further; from the original "likely" (66%) it went up to "very likely" (90%), and it will go up again." And yes, indeed, in Stockholm, nearly seven years later in September 2013, it became "extremely likely" (95%).

One issue with the IPCC methodology of likelihood statements has already been addressed: The methodological unreliability of models has been used to "downgrade" likelihood statements without saying so. Another issue, which is related to insufficient transparency of expert judgment in the IPCC, is that there is hardly any reflection on the nature of expert judgment. "Very likely" means more than 90% chance that a particular statement is true. But what does that really mean? What do these probabilities mean? How reflexive is the IPCC about what is actually happening, and what is behind these statements? The "90%" only means that the few authors who have been selected to do the assessment in a particular chapter have somehow reached this collective expert judgment. Nothing more and nothing less. It carries a lot of weight, because these authors have had the scientific training, acquired the relevant skills, and have a lot of experience in their scientific practices—they bring all these things to the table. These lead authors are the experts. We choose them for that expertise. Then, other experts are asked to thoroughly review their statements. The lead authors, however, in the end, when they write down their conclusions, get rid of any reference to "expert judgment." Suddenly their conclusions are made to flow directly from the underlying science. "It is not us." I find it incredible!

Twice we have had to intervene as the Dutch government delegation asked to make the Summary for Policymakers more explicit about expert judgment. In Paris in 2007, for example, the authors, when defining their uncertain terminology, referred in the final draft to the "assessed likelihood of an outcome or a result." We added "using expert judgment" to that phrase. In Stockholm in 2013, the same problem arose with the definition of "probabilities": "Probabilistic estimates of quantified measures of uncertainty in a finding are based on statistical analysis of observations or model results, or expert judgment." We looked at it and saw that it was going in the wrong direction. We thus changed "or expert judgment" into "and expert judgment." I think this is important. It is worrisome that scientists who act as science advisers are often not able to reflexively say what they are doing.

Questions on how expert judgment can be reflected in the IPCC are intertwined with questions on how science and politics relate in the IPCC. I would like to frame IPCC assessments as social constructs with elements from both science and politics. Thus both types of values are in play: values both intrinsic and extrinsic to science. How successful is the IPCC? Well, critics would say they are too successful in terms of connecting with policy, and unsuccessful in connecting with science. That issue is what I studied for the Third Assessment Report (published in 2001), to address criticism in the US Senate testimony by Dick Lindzen that the IPCC would not be open enough to skeptics.

In addition to too little reflexivity in the IPCC, I also found that the criticism of lack of openness to skeptics was incorrect. For the report that I studied (I took the chapter on attribution of climate change to human influences), I looked at all the comments there were submitted for that chapter in all the review rounds. I looked at all of the responses to those comments, and all of the review-editor comments to the responses, and discovered that there were a lot of critical comments, many of which had led to improvements in the text in terms of more inclusion of uncertainties and better language.[10] So, I do think that skeptics (taken in a broad sense, i.e., including not only the "typical" climate skeptics but also people who for good reasons are critical of climate modeling) play a constructive role in the IPCC process. The final outcome is a policy-relevant assessment. It is not, however, the scientific consensus with full certainty, and thus it should not be framed in this way. Of course, the IPCC can still further improve its communication of uncertainty, be more transparent, and explain where the expert judgments come from, to connect with what Heather also emphasizes in her lecture. And I think the IPCC could be more reflexive about what is actually happening in these plenaries. They are all closed. Why? Include a webcast, for instance. There is no reason not to do that.

I conclude with four lessons that I took from my 14 years of being a science adviser:[11]

1. **Explicit reflection on uncertainty and values.** Take "normal science" seriously, but also organize reflection on its uncertainties and value-ladenness.

I have bought into the discourse of post-normal science, while I do agree with Heather that there never was a period where there was no post-normal science. So "post-" should perhaps read "extra-": "extra-normal science." With "normal science," I really mean those proceedings where it is the scientific community that is doing whatever they are doing: modeling, publishing, peer reviewing, etc. So when I say that we need to open up look at ways to bring out the different epistemic and nonepistemic values in this discussion, I mean that we need to organize reflection on uncertainty and value-ladenness within normal science as well, without throwing it away. So don't throw away the baby (post-normal science) with the bathwater (a form of scientism that does not sufficiently reflect the presence of uncertainty and ignorance in science)! Hence, I do not buy into very simplistic readings of post-normal science.

2. **Addressing methodological and public reliability.** Alongside the *statistical reliability* of results (expressed in terms of probability), devote due attention to their *methodological reliability* (expressed in terms of strengths and weaknesses) and their *public reliability* (expressed as the degree of public confidence in the scientists who produce them).

As I have already belabored in this essay, do not focus only on statistics; also focus on qualitative dimensions of reliability.

3. **Extended peer review.** Involve a larger group of specialists and nonspecialists who hold different values in monitoring the quality of scientific assessments.

"Extended peer review," which also comes out of this literature of post-normal science, concerns the ways in which one can engage a wide group of people who can provide comment and are sensible enough so they can be processed and responded to, for instance, in the IPCC. Everybody—on the basis of a very minimal claim to expertise—can sign up to be an expert reviewer of the IPCC and can submit comments. It is very important that not only is a very small group of climate modelers, for instance, providing comments on the climate modeling chapter, but so too are neighboring disciplines and people who work for Greenpeace, for example. They all have a stake, as well as very valuable contributions to bring, because they can highlight particular risks to the climate that may not have become mainstream yet in the scientific community.

4. **Acknowledging social complexity.** Be wary of accepting the conclusions of actors and practitioners at face value; try to delve deeper through the layers of complexity by means of narrative methods.

The final point—looking at deeper dimensions and different things that are happening at the same time—is related to the notion of "social complexity." Scientists often have a self-image (overly rationalized) of what they are doing and the country delegates have a self-image (again overly rationalized) of what they are doing, and these self-images are too simplistic in what they hold, because they do not reflect the complexity of the way different types of values (epistemic and nonepistemic) are interwoven in practices. In terms of how to understand this, it is important to delve deeper. The big question still remains: Are there improvements that we can suggest to this mess? It is a mess, but already a good and interesting mess.

Notes

[1] I have been a Dutch government delegate to the IPCC from 2001 to 2014.

[2] See, e.g, Arthur C. Petersen, Albert Cath, Maria Hage, Eva Kunseler, and Jeroen P. van der Sluijs, "Post-Normal Science in Practice at the Netherlands Environmental Assessment Agency," *Science, Technology, & Human Values* 36, no. 3(2011): 362–388.

[3] Cf. L. A. Meyer and A. C. Petersen, eds., *Assessing an IPCC Assessment: An Analysis of Statements on Projected Regional Impacts in the 2007 Report*, PBL Report 500216002 (The Hague and Bilthoven, Netherlands: PBL Netherlands Environmental Assessment Agency, 2010).

[4] IPCC, *Climate Change 2001: The Scientific Basis. Contribution of Working Group I to the Third Assessment Report of the Intergovernmental Panel on Climate Change* (Cambridge, UK: Cambridge University Press, 2001), 14.

[5] See Arthur C. Petersen, *Simulating Nature: A Philosophical Study of Computer-Model Uncertainties and Their Role in Climate Science and Policy Advice*, 2nd ed. (Boca Raton, FL: CRC Press, 2012), 145.

[6] For a transcript of what transpired at that contact group meeting, see Arthur C. Petersen, *Simulating Nature: A Philosophical Study of Computer-Model Uncertainties and Their Role in Climate Science and Policy Advice*, 2nd ed. (Boca Raton, FL: CRC Press, 2012), 191–197.

[7] Ninety-five percent or "extremely likely" was only added to the methodology in the most recent assessment round.

[8] Arthur C. Petersen, *Simulating Nature: A Philosophical Study of Computer-Model Uncertainties and Their Role in Climate Science and Policy Advice*, 2nd ed. (Boca Raton, FL: CRC Press, 2012); Leonard A. Smith and Arthur C. Petersen, "Variations on Reliability: Connecting Climate Predictions to Climate Policy," in *Error and Uncertainty in Scientific Practice*, Marcel Boumans, Giora Hon, and Arthur C. Petersen, eds. (London, UK: Pickering & Chatto, 2014), 137.

[9] Matthijs Kouw and Arthur Petersen, "Diplomacy in Action: Latourian Politics and the Intergovernmental Panel on Climate Change," *Science & Technology Studies* 31 (2018): 52–68.

[10] Arthur C. Petersen, *Simulating Nature: A Philosophical Study of Computer-Model Uncertainties and Their Role in Climate Science and Policy Advice*, 2nd ed. (Boca Raton, FL: CRC Press, 2012).

[11] Arthur C. Petersen, "The Ethos of Scientific Advice: A Pragmatist Approach to Uncertainty and Ignorance in Science and Public Policy," in *Building Bridges: Connecting Science, Technology and Philosophy – Essays presented to Hans Radder*, Henk de Regt and Chunglin Kwa, eds. (Amsterdam, Netherlands: VU University Press, 2014), 53.

Commentary on Lecture 2

EXPERTISE AND ACCOUNTABILITY

Torsten Wilholt

In her second Descartes lecture, Heather Douglas addresses issues that are both politically extremely pressing—some might say depressing—and philosophically highly interesting. As is characteristic of most of her work, epistemological concerns, on the one hand, and questions of ethics and political philosophy, on the other, are inextricably intertwined. Two main topics stand out from the wealth of helpful and innovative ideas presented in the second lecture: the accountability of experts, and policies of research funding. In the following remarks, I comment on each of these in turn.

The Accountability of Experts

Douglas approaches the topic of the accountability of experts by making some concrete claims about how it ought to be realized. Experts should be accountable for the accuracy of their claims to their peers from their respective research communities. And they should be accountable to the citizenry for the value judgments they inevitably make in weighing the uncertainties involved in their expertise. Last but not least, for actually making such methodological decisions that are in accordance with the value outlook they explicitly endorse, experts should also be accountable to their home research community.

From this range of ideas, I will focus on experts' accountability to the citizenry, since it seems to contain (potentially) the most controversial suggestions. Douglas proposes that in the case of formal policy advice, accountability should be achieved by means of a pluralism of perspectives within expert committees, and by means of experts expressing their value judgments explicitly. Both suggested measures are arguably commendable on independent grounds. However, it is not obvious that the value of pluralism in expert committees derives from its contribution to making expertise accountable to the citizenry. In so far as pluralism contributes to accountability at all, its immediate effect rather seems to be to make individual experts sitting on the committee accountable to *each other*. To conceptually extend this to an accountability to the citizenry appears to rest on the implicit assumption that, by virtue of possessing different value outlooks, the experts on the committee act as *representatives* of the citizens whose value outlooks they share. However, this is a tenuous kind of representation, lacking as it does in, well, accountability. Typically, experts do not *answer* to the citizens whose value outlooks they "represent"; they cannot be recalled or deselected by them.[1]

A similar point arises with respect to accountability in informal policy advising. Douglas suggests that it is established by virtue of the close relationship with the elected official whom the advisor advises. But whatever accountability to the citizenry the advisee is subject to does not automatically transfer to the advisor via this relationship. When the official fails to be reelected, the advisor may lose her privileged access to the corridors of power. But in the vast majority of cases, the advisor has her main occupation elsewhere, so her stakes in the reelection are considerably less high as compared to the advisee.

Note that by expressing my own doubts about whether the arrangements proposed by Douglas really implement accountability of experts to citizens, I do not mean to suggest that we ought to adopt more stringent measures that would do so. I do not think that we should make experts answerable to a constituency. It would compromise their freedom of judgment. Douglas' systematic treatment of expert accountability provides us with an excellent way of articulating the reason why: Experts should only be accountable to the citizenry for the values they employ in weighing uncertainties, not for the accuracy of their claims, for which they should only be accountable to their peers. If experts were elected by citizens, it would be impossible to separate the two kinds of accountability from each other. A case in point is the situation that developed after the Flint water crisis had reached its climax. Marc Edwards, the Virginia Tech engineer who was instrumental in proving the dangerously unhealthy quality of Flint's tap water, was in a certain sense a policy advisor who was selected by the citizenry, since he only got involved in the matter after activists had called upon him to help them. However, after unsafe levels of lead in the water had been exposed with Edwards' help and Flint had switched back to Detroit water as a consequence, Flint citizens started turning away from Edwards as he no longer seemed to be delivering the kind of information that they wanted to hear.[2] Many citizens were subjectively convinced that the water quality was not improving after the switch, a judgement that Edwards was not able to confirm. In their search for alternative water expertise, some citizens soon started to turn to a Hollywood actor and a businessman who claimed to have invented a novel sponge product for more accurate water testing. These two were alleging that Flint water was still unsafe, even for bathing, due to contaminants other than lead. (The businessman has since backed down on this claim.)

This story is apt to reinforce a point that is implied by Douglas' original discussion, namely that the desired accountability of experts to the citizenry for the value judgments they employ must be kept separate from an accountability for the *content* of their expertise, which could easily undermine its credibility and render it pointless. The need to keep the two apart also places limitations on the extent to which even the desirable kind of accountability to the citizenry can be effectively put into practice. I do agree, however, that transparency and open communication about value judgments, as well as the representation of a plurality of perspectives in expert committees, would take us in the right direction.

How far would they take us in our efforts to let experts be "on tap, but not on top"? At the core, there is the problem, in Philip Kitcher's words, "to decide how to integrate the plausible idea that, with respect to some issues, some people know more than others, with a commitment to democratic ideals and principles."[3]

Douglas shows that we can get some leverage on this problem by focusing on accountability. Accountability certainly is a central and indispensable instrument of the democratic process. However, accountability is not the beginning and the end of democracy—not, that is, unless we subscribe to a minimalist, Schumpeterian conception by which democracy is limited to negative control and the sole role of the people is to produce a government.[4] To be sure, Douglas does *not* commit herself to a minimalist view of democracy, nor do I have any reason to believe that she would want to. But for what follows, I am going to rely on the premise that we ought to aspire to a much richer form of democracy than that.[5] And so, without going into any of the fine points of the debates within democratic theory, I will at least briefly sketch one reason why negative control is insufficient.

Sometimes, there are solutions to political problems that are, in a robust sense, good solutions for the community as a whole—that is, they are not just combinations of or compromises between those solutions that individuals would regard as good when judged from their particular perspective. Simple game theoretical considerations illustrate that this is at least possible. It is not to be expected that we will be able to actualize or even identify these good solutions if all that each of us does is state their individual interests and bargain for the deal that serves them best. The democratic process must be organized in such a way as to encourage the search for a shared vision of the common good and to enable deliberative processes that identify solutions that best serve the common good. To the degree that citizens can participate in this process of public reasoning, it is to be expected that lines of reasoning in favor of and against the different political options that are addressing the participating citizens' viewpoints and concerns are going to be produced in the course of the debate. This sort of public justification of political options is one of the main features of democratic legitimacy that set it apart from other forms of government.[6]

All this means that democracy requires processes of public reasoning that are geared toward discovering, revising, and developing a shared conception of the common good, and opportunities for citizens to participate in these processes. Obviously, knowledge plays a great role in this view of democracy. Without knowledge, it is impossible to participate effectively in such a way as to bring one's values and interests to bear.

In consequence of this, expert knowledge must first and foremost be knowledge that is directed at informing the citizenry. It should feed into processes of public deliberation. The more it is delivered into the hands of politicians behind closed doors, the less it contributes to democratic legitimacy and to the power of the people to govern themselves.

It might be objected that all of this adds up to a far too demanding or even utopian conception of democracy that has little to do with the messy realities of our political communities. But I maintain that it has a very concrete bearing on some of the most pressing concerns with present-day political life in liberal democracies, especially with regard to the role of science and expertise.

The credibility of experts in the context of democratic politics is often diagnosed to be in crisis. An incisive moment came when Michael Gove, then Lord Chancellor and one of the leaders of the Leave.EU campaign in the British EU membership referendum, used the following language to cast aside the many predictions from the International Monetary Fund, the Organization for Economic Co-operation and Development, and the European Central Bank—that is, the IMF, the OECD, and the ECB—in addition to other organizations, which outlined the bleak economic prospects awaiting the UK when it breaks away from the European Union: "I think the people of this country have had enough of experts [...] from organisations with acronyms saying that they know what is best and getting it consistently wrong [...]."[7] What is noticeable about this remark is perhaps not so much that Gove would make it, but that he and the Leave.EU campaign have fared so well with this stance, particularly given the glaring untruths on which some of the campaign's arguments were based. The parallels to Donald Trump's US presidential election campaign have been noted many times, and commentators have used the label "post-truth politics" to describe the indifference to truth that seems to underlie these phenomena.[8]

It is easy to get frustrated and angry about post-truth politicians and their supporters, but one should resist the urge to attribute their success to an epidemic of ignorance and stupidity. Much more helpful is Douglas' perceptive observation that the contempt for science and experts that is spreading amongst small-government conservatives is at least in part explainable by the fact that, in many contexts, scientific knowledge has a great potential for shifting the purview of the public, and hence of government. I think this is an important root of the problem, but perhaps not yet an explanation of its full extent.

Across Western democracies, governments are increasing their use of the tactics of presenting "necessary" policies to the citizenry that are portrayed as immediate results of factual constraints. Expert advice has become crucial as a tool for the legitimization of policies along these lines. Public deliberation is thus often short-circuited, and there are no opportunities for participation in an open-ended process of public reasoning. In parts of the citizenry, this contributes to a feeling of being governed by distant and aloof elites. The growing resentment of the politics of factual constraints is transferred to the experts. They are starting to be regarded as implicated in networks of power and as serving the interests of elites. This undermines their credibility.

If knowledge is a prerequisite of effective participation in the democratic process, and if the required knowledge now often specifically includes the kind of highly specialized information only professional specialists can provide, then a crisis of the credibility of experts is harmful to citizens and harmful to democracy. Therefore, one main challenge seems to be to restore and maintain the credibility of experts in light of their involvement in politics. While accountability and credibility are of course related, and the measures proposed by Douglas for strengthening accountability would likely also benefit credibility; credibility implies a broader range of issues than accountability. In the epistemic realm, trustworthiness cannot be ensured by negative control.

If we widen the perspective to other aspects of credibility than accountability, we will have to consider a variety of other factors as well. If experts are regarded as *integrated* in elite networks, it is sometimes because they are. Changes in science policy, such as the intentional creation of incentives for university researchers to profit financially from their discoveries by means of patents and spin-offs, have changed what it means to be successful in the sciences to also include—at least in some fields—financial success. Maintaining good relations to people of influence outside academia has become a common aspect of many academics' work life—it can certainly be useful and perhaps sometimes necessary in order to obtain funding. In many cases, scientists no longer work in the kind of institution that guarantees their independence and embodies it in a way that inspires trust. The relation between different forms of institutional and individual independence and the credibility of expertise is a subject that should be of central concern to philosophers and sociologists of science.

Another area of concern is the mechanisms and institutions by which scientific expertise is delivered. In some technical matters as well as in urgent cases, consultations between experts and government officials behind closed doors are clearly appropriate, but they should remain the exception. Ideally, relevant scientific expertise should be publicly available early on and should feed into procedures of deliberation. Engaging publicly and at an early stage with developments of potential relevance to citizens may help scientists to regain public trust if they manage to convey that they intend to provide input for public debate rather than to pass it by.

Lastly, the credibility of scientists for citizens is based, to a large degree, on their accountability to each other. The present state of this type of accountability also leaves room for improvement, as evidenced by the frequent calls for more full and transparent sharing of data, analytic code, and clinical study reports, for full and detailed pretrial registration and for post-publication peer review.[9] Some of the suggested improvements also create opportunities for a transparent discussion of value-laden issues and their bearing on the management of inductive risks. If post-publication peer review created a forum to also discuss questions of inductive risk more openly, this could make these questions more transparent for citizens without forcing scientists to cater to public opinion.

Funding and Public Interest Science

I conclude with a few remarks on science funding, the other thematic focus of Douglas' second Descartes lecture. When it comes to contemporary trends such as the economization of science, many scientists are eager to point out that curiosity-driven basic research must not be allowed to fall behind. Douglas rightly reminds us that besides basic science and science in the private interest, there is also public-interest science that deserves our attention. She proposes oversight committees in order to foster research in public-interest science. At the same time, she also expresses reservations against too much central planning.

Of the many reasons to doubt the value of central planning in science, I would like to briefly unpack one:[10] Central planning can never be as good at utilizing local knowledge and mobilizing individual creativity as decentralized, self-organizing processes are. Here, by local knowledge I mean the kind of knowledge that individual researchers have about their own talents and skills, about existing technical equipment and other resources available to them, as well as about contacts and networks that can be activated for research on a particular approach. These are the kinds of resources that individual scientists draw on when looking for a research project that maximizes their chances of success and recognition. Since local knowledge is constantly in motion and often challenging to communicate, it is difficult to think of any efficient way of utilizing it for shaping the research agenda that does not rely on decentralized individual initiative and competition.

To a certain extent, I fear that this would also affect the work of any committee tasked with identifying the gaps in public-interest science that need to be filled. Identifying such a gap requires two things at once: identifying the public need for a certain kind of knowledge, and realizing that this kind of knowledge is within reach from the perspective of present-day science. It is the second of these that will often require local knowledge and creativity, especially if cutting-edge research is involved. "Research," as Peter Medawar once put it, "is surely the art of the soluble."[11] Seeing which problems are now well enough defined to allow targeting with the cutting-edge methods of research will often presuppose the kind of gestural and situational knowledge that exists dispersed over the members of a research community and is not easy to harvest for a central committee.

A better hope for public-interest science may lie in creating and improving incentives for this kind of work. Many scientists express an intrinsic interest to do work that benefits the common good. Contributions to public-interest science that give extra weight to tenure and promotion decisions, awarding of prizes, or the selection of members of learned societies can be expected to amplify and support this interest. At the moment, this would constitute something like a counter-current to the official institutional cultures that are still promoted within many organizations and that typically emphasize bringing science to the market. Universities in particular still seem to be stuck in this 1980s vision. Several universities have adopted the label "the entrepreneurial university." The "public-interest university" seems to be suspiciously absent from the lexicon of public relations for universities.[12] Institutional cultures affect scientists' choices, and therefore the valorization of public-interest science could be key to improving the imbalance that Douglas rightly points out. Cultural changes work slowly and unpredictably, but I see more reason for optimism that they will have a positive effect than in the case of oversight committees.

Notes

[1] See Mark B. Brown, *Science in Democracy: Expertise, Institutions, and Representation* (Cambridge, MA: MIT Press, 2009), for a discussion of the topic of representation in science and democracy.

[2] Donovan Hohn, "Flint's Water Crisis and the 'Troublemaker' Scientist," *The New York Times Magazine* (Aug. 16, 2016).

[3] Philip Kitcher, *Science in a Democratic Society* (New York, NY: Prometheus Books, 2011), 20.

[4] Cf. Joseph Schumpeter, *Capitalism, Socialism, and Democracy* (New York, NY: Allen & Unwin, 1976), Chapter XXII.

[5] The richer account of democracy that I will presuppose is roughly based on deliberative ideals, such as have been proposed by Jürgen Habermas in *Between Facts and Norms*, trans. W. Rehg (Cambridge, MA: MIT Press, 1996); John Rawls in *Political Liberalism*, expanded edition (New York, NY: Columbia University Press, 2005); and others.

[6] For a discussion of democratic theory that develops a similar line of reasoning, see Joshua Cohen, "Deliberation and Democratic Legitimacy," in *The Good Polity: Normative Analysis of the State*, Alan Hamlin and Philip Pettit, eds. (Oxford, UK: Blackwell, 1989), 17.

[7] Sky News, "Faisal Islam Interview with Michael Gove" (June 3, 2016, 8:00 pm), available online: https://www.youtube.com/watch?v=GGgiGtJk7MA.

[8] See, e.g., William Davies, "The Age of Post-Truth Politics," *The New York Times* (August 24, 2016).

[9] Ben Goldacre and Tracey Brown, "Fixing Flaws in Science Must Be Professionalized," *Journal of Clinical Epidemiology* 70 (2016): 267–269; Jaime A. Teixeira da Silva and Judit Dobránszki, "Problems with Traditional Science Publishing and Finding a Wider Niche for Post-Publication Peer Review," *Accountability in Research* 22, no. 1 (2015): 22–40.

[10] Cf. Torsten Wilholt, "Scientific Freedom: Its Grounds and Their Limitations," *Studies in History and Philosophy of Science Part A* 41, no. 2 (2010): 174–181.

[11] Peter B. Medawar *The Art of the Soluble* (London, UK: Methuen, 1967), 87.

[12] Ironically, the universities tagging themselves as entrepreneurial typically seem to be European public universities, funded to an overwhelming extent by taxpayers' money and thus bound to the public interest on pain of the legitimacy of their very existence. The great private American universities (which are truly entrepreneurial entities) being dependent on benefactors that want to be assured of the philanthropic character of their donations, are much less likely to showcase the entrepreneurial nature of their institutions.

Lecture 3

SCIENCE COMMUNICATION: BEYOND THE DEFICIT MODEL

Heather Douglas

Introduction

The value-saturation of science both poses challenges and opens opportunities for the interactions of science and democratic governance, of scientists and democratic publics. In the previous lecture, I focused on government interactions with science and scientists, addressing both science advisors and science funding. The discussion of science-policy interfaces was far from complete, and filling in further details and examining a fuller range of science-policy interfaces must await future work. Yet the public was largely ignored in that lecture, interacting with science primarily through the election of and evaluation of public officials, based on those officials' values and responses to science advice. This is, of course, not the central way in which the public and the scientific community interact. There is usually a more direct connection between these two groups, a connection which is the purview of science communication.

Science communication is a broad field, encompassing media reports and press conferences, science museums, television and film presentations of science, and formal science education. Things have come a long way from the days when Carl Sagan was vilified within the scientific community for his work on *Cosmos*;[1] scientists now recognize the importance of science communication, and work to be better at it. There are science communication training modules for scientists; there are awards within scientific societies for those who do science communication well (e.g., the AAAS Award for Public Engagement with Science). The scientific community clearly recognizes that if the public is to both understand and benefit from the work scientists do, science communication must be done well.

But scientists have also been frustrated by the impact of their efforts in this area. Although science is one of the most trusted institutions in contemporary societies—rating higher than the media, organized religion, and the government in most polls—the public (or more precisely some segments of the public) sometimes do not accept or believe what scientists are trying to communicate. Topics on which a substantial part of the public is skeptical about a strong scientific consensus range from climate change and evolution to GMO safety and vaccines. Recent polls of scientists show that most scientists (84% of them) place the blame for this rejection of scientific claims on the public's lack of knowledge about science.[2]

Scientific literacy surveys further support this view of where the problem lies. When taking such surveys—which usually entails answering simple questions (e.g., true/false format) about what scientists think of as "basic facts"—the public does poorly. No nation has even half its adult population passing these basic scientific literacy tests. While there appear to be gains in some countries in recent years on such baseline science literacy, it is still abysmal, and scientists routinely blame this illiteracy for the failure of the public to accept scientific findings. After all, if you don't understand basic scientific information, how can you decide what is scientifically sound or not? This is the so-called deficit model of science communication—that the public fails to accept or believe science because it simply doesn't know enough science.

In this lecture, I will *not* argue that scientists are wholly misguided to think that the public's lack of knowledge about science is a problem. Rather I will argue that what scientists normally think of as the requisite content of this knowledge is off target, that what the public needs to know about science is not the usual package of a set of facts, of bits of knowledge without which the public will have no understanding of what scientists are talking about. Rather, I will suggest that what is important for the public to understand about science is the nature of scientific reasoning and practice, and that factual scientific knowledge is secondary to this understanding.

Further, the understanding of science we need the public to have includes not just the role of evidence but the role of values in scientific practice. With a public informed by such a picture, scientists will need to commit to more of a dialogue with the public rather than the one-way transmission of bits of information. Science communication, when done properly, becomes a two-way discourse rather than a one-way stream. I will describe some of the mechanisms and institutional settings that can assist us with such dialogues toward the end of this lecture.

I will start by describing the baseline view that the public is problematically scientifically illiterate and that this is the reason for the discrepancies between what the public believes and what scientists believe (the classic deficit model). Empirical challenges to the deficit model have arisen in recent years, undermining the validity of the deficit model. These are important challenges, but they have not come with appropriate frameworks for understanding what we *should* be trying to achieve in science communication. I will argue that these newer approaches for understanding science communication fall short. Part of the problem is that the current theories of science communication do not have a rich enough account of the proper roles for values in science, and thus, view any influence of values on the acceptance of scientific claims as not properly rational. I argued in the first lecture that this is a mistake, that there is a very important way that values can legitimately influence the acceptance and rejection of scientific claims—and this applies to both scientists and the nonexpert public. Acknowledging this role means science is no longer a value-free resource for public debate. It also opens up the possibilities for debate and discourse along productive lines. Rather than stalemates of ideological clashes, we can see how understanding the intertwining of values and evidence in science provides ways to disagree rationally and to discuss our disagreements without disparaging our opponents. Science communication becomes a different endeavor with this conception, far more interactive in character. But in order for such interactivity to work in practice, the public will need to understand the nature of science better. Science education should focus on this better understanding of the practice of science, producing a basis for more interactive and ultimately more robust science communication.

Scientific Literacy and the Deficit Model's Demise

The perceived crisis of scientific illiteracy in the US dates back at least to the late 1950s, when the country was in the grip of concern over the technological prowess of the Soviet Union, exemplified by the successful launch of Sputnik in 1957. Six months before the launch of Sputnik, the National Association of Science Writers conducted a survey of the US population regarding their perception and understanding of science.[3] Although the survey focused on public interest in science news stories, it also included some fact-based questions that attempted to assess how much the public understood about particular scientific topics of the time: radioactive fallout, fluoridation, the polio vaccine, and satellites (a topic on which the public would have done better after Sputnik). Less than 20% of those surveyed had accurate understandings of all four topics; most Americans understood two or less.[4] Similarly low levels of literacy were found in other surveys of the period.[5] By the 1970s, more systematic efforts to measure scientific literacy, and discussions of what should constitute scientific literacy, had begun.

Debates over how to conceive of scientific literacy were tied in the 1970s to the idea that scientific literacy was essential to democratic policy making. The idea of a civic scientific literacy—"a level of understanding of scientific terms and constructs sufficient to read a daily newspaper or magazine and to understand the essence of competing arguments on a given dispute or controversy"[6]—was made distinct from the ability to write clearly about science. It was this kind of literacy that seemed essential to society. For example, physicist Benjamin Shen argued in 1975 that what we needed to worry about was an adequate understanding of basic scientific ideas, and that this "civic science literacy [was] a cornerstone of informed public policy."[7] Today, this sense of science literacy is still seen as essential for democracy, even for civilization itself. For example, a recent Smithsonian report declares: "Scientific literacy is an urgent and important issue. Why should we care? The answer is simple: Our way of life and our survival are at stake."[8]

As this conception of scientific literacy emerged, the initial challenge was how best to measure this kind of literacy, given that the scientific controversies in the news changed over the years. Beginning in the late 1970s and with support from the National Science Foundation, Jon

Miller worked to provide a "durable" measure of scientific literacy, one that could weather changes in the topics of interest of the day. The kinds of questions asked, what are now called "science education/literacy indicators," are simple true/false questions that assess whether the public grasps basic "facts."[9] What has become a standard technique, though, does not reflect Miller's full view. In his work on generating measures of civic scientific literacy, he embraced a three-part conception of what literacy should entail: 1) a grasp of basic scientific concepts (i.e., scientific facts), 2) an understanding of scientific inquiry, and 3) an understanding of the impact on science and society.[10] This is indeed a useful tripartite goal for assessing scientific literacy, but in practice, the first part usually swamps the second and third parts. This is probably because asking about facts provides the most obvious fodder for ease of testing and evaluation. Standard science literacy tests usually contain a dozen or so true/false questions about scientific facts (e.g., "The center of the earth is hot." or "An atom is larger than an electron.") and a couple of open-ended questions about scientific inquiry, such as "Describe what an experiment is" Not surprisingly, the questions about facts come to dominate discussions of scientific literacy, as it is over these questions that statistics can readily be gathered.

With the regularly dismal showing of democratic publics on these kinds of tests, many scientists have come to blame the public's ineptitude on scientific literacy tests for their failure to accept the authority of expertise regarding issues of public import. [11] This is the "deficit model," the idea that the problem of the public contesting or rejecting scientific expertise arises primarily because of the scientific illiteracy of the public.

Since 1990, the deficit model has been subjected to some potent critiques. Brian Wynne, in his classic 1992 paper "Misunderstood misunderstandings," showed how the issue with science communication in the case he examined was not a lack of knowledge by the "lay public," but a lack of knowledge by the experts regarding the actual local conditions and practices to which their expertise was supposed to apply. The experts simply didn't know enough about sheep farming or local soils to be able to construct appropriate field trials and make accurate predictions crucial for ensuring the safety of agricultural practices and products in the region after the Chernobyl accident.[12]

More recently, political scientists who examined population-level trends in skepticism about various contested scientific claims empirically disconfirmed the deficit model. These scientists tested whether people who do more poorly on fact-based scientific literacy tests are those less accepting of scientific claims, and found the *reverse* to be true.[13] Instead, they found that those most skeptical of key contested claims do better on the scientific literacy tests. What drives disagreement seems to have more to do with ideological commitments or value frameworks than whether someone has a grasp of scientific facts. The higher the education level, the more some members of the public feel competent to challenge expertise. For example, among social conservatives, the more people know about science, the more skeptical they are about climate change. And it is often the more highly educated members of the public who are the strongest critics of vaccine policy. Therefore, the deficit model does not appear to hold up empirically.

This disparity might be because the scientific literacy tests are too simplistic. For example, knowing that the earth goes around the sun once each year is simply not enough for understanding climate change. One can be skeptical that knowing whether electrons are smaller than atoms is even relevant to the issue.[14] Perhaps the problem is that a little bit of knowledge is dangerous, that having a bit of familiarity with an area of science allows one to feel freer to disagree with experts. But we do not want a more ignorant, and thus more pliable, public. Nor can we gain expertise in everything—some division of cognitive labor is needed. So there will always be areas where we have only a bit of knowledge. Rather than increasing expertise and familiarity with all the areas of publicly relevant science, or aiming for *reduced* scientific literacy (!), we need another path.

The place to start is looking at the causal factors that drive disagreement with scientific expertise (particularly with a consensus of scientific experts), aside from scientific literacy. Recent work suggests that disputing the experts has a lot to do with value commitments and worldviews.

In Dan Kahan's work, for example, he has championed the idea that ideological worldviews determine much of what the public believes. Within his theoretical framework, he has developed tests to sort the population by whether one is egalitarian (tending to prefer social arrangements of presumed equality) or hierarchical (tending to prefer hierarchical arrangements such as those inside the military or private business), and whether one is individualist (valuing the needs of humans as individuals first) or communitarian (valuing the needs of communities first).[15] He has found that rather than distributing themselves among the four quadrants, people (at least in the US) tend to cluster into egalitarian communitarians or hierarchical individualists. Kahan finds that depending on how one sorts into these two categories (based on his personality tests), one is more likely to accept or reject certain scientific claims. For example, egalitarian communitarians are more likely to accept scientific claims about anthropogenic climate change, whereas hierarchical individualists are less likely. Even more devastating for the deficit model, if a person is a hierarchical individualist, the more that person knows about the science, the *more* likely they are to reject the consensus on climate change.

Other social scientists such as Matthew Nisbet have examined particular acts of communication and have noticed the impact of framing on the acceptance or rejection of scientific claims. In these studies,[16] social and ethical values, and how they are drawn upon in a piece of science communication, are shown to have a substantial impact on the uptake of particular claims. These social scientists argue for reflective science communication strategies that package, or frame, scientific information in ways that will reach audiences. Doing so might require multiple packaging of the same information for different audiences.

Both of these recent approaches build from insights on motivated reasoning, an idea originating from Ziva Kunda's classic 1990 paper, "The Case for Motivated Reasoning." Kunda surveyed wide swaths of psychological research to show that "directional reasoning" was an endemic aspect of human reasoning, that such directional reasoning was more concerned with reaching a particular conclusion than with getting to the truth (she called truth directness as being motivated by accuracy), and that wishes, desires, and values affected both conclusions and reasoning processes. Kunda posited that our motives shaped our memory search (leading to selectivity in the empirical evidence on which we draw) and our belief-construction processes.[17] While much of the evidence Kunda discussed drew from research about people's beliefs about themselves and others (notoriously complex subjects), some studies concerned people's beliefs regarding factual statements about the world (such as the accuracy of medical tests or the dangers of consuming caffeine) or people's use of statistical rules (such as an awareness of the importance of base rates).[18]

Although Kunda argued that directional reasoning is a pervasive aspect of human thought, she also noted that directional reasoning did not mean that one could believe whatever one chose. Evidence still constrained such reasoning, and without some evidence or reason to support one's views, even directional reasoning failed to produce the desired end. "People attempt to be rational: They will believe undesirable evidence if they cannot refute it, but they will refute it if they can."[19]

One can view the work of current social scientists like Kahan and Nisbet as demonstrating the pervasive use of motivated reasoning regarding publicly controversial scientific topics. But this just seems to make our situation with respect to science communication worse.

On the one hand, openly saying that members of the public support certain views because they accept the science that fits with their worldviews or their values, and they reject what does not, is disparaging of the public.

Consider Kahan et al.'s discussion of their work (2012):

> *Our findings could be viewed as evidence of how remarkably well-equipped ordinary individuals are to discern which stances towards scientific information secure their personal interests.... For the ordinary individual, the most consequential effect of his beliefs about climate change is likely to be on his relations with his peers. A hierarchical individualist who expresses anxiety about climate change might well be shunned by his co-workers at an oil refinery in Oklahoma City. A similar fate will probably befall the egalitarian communitarian English professor who reveals to colleagues in Boston that she thinks the scientific consensus on climate change is a hoax.*[20]

Kahan et al. go on to argue that because each individual's actions in mitigating climate change is likely to have negligible impact, it makes sense that getting along with one's peers overrides any concern one might have about climate change for either group. Yet to tell members of the public that this is the reason they accept or reject a claim, because it fits with their ideological worldview, because it is what their neighbors and friends accept, is insulting. (Imagine if I told you that was the reason you held the beliefs you did—that while you might think you had other reasons, the real reason was because you want to fit in with your friends and colleagues.) Although this kind of motivated reasoning can be construed as rational in the sense of pursuing one's self-interests, it falls short of the rationality that one would want on public display or in a space of public reasoning.

Openly describing and discussing the public's views in this way is thus problematic. We could, on the other hand, use an awareness of values in science communication to tailor communication regarding science to increase uptake of the science and not say openly that this is what is occurring. We could use the theories to frame communications to reach recalcitrant publics. But this just seems manipulative. And indeed, social scientists working with framing theory raise such concerns. As Nisbet and Scheufele note:

> *Public communication and engagement should not be conceived of as simply a way to "sell" the public on the importance of science or to persuade the public to view scientific debates as scientists and their allies do. To apply sophisticated approaches such as framing or deliberative forums to achieve these ends falls back into the trap of deficit model thinking and undermines longer-term efforts at building trust, relationships, and participation across segments of the public.*[21]

It seems we are stuck. It does not seem plausible that we can use such theories openly, as that seems insulting and disparaging of the public. Telling someone you are tailoring the message to their worldview and/or values because they will more likely accept it using motivated reasoning would not work, at least in the long run. And to use this work without acknowledgement is problematic as well. We don't want to just manipulate the public into accepting scientific consensus. Doing so would likely be self-defeating, as the public would probably notice that different science communication messages are uniquely tailored, and become suspicious of such communication efforts.

The approaches described here seem to make the disagreements regarding politically relevant science permanently intractable. If people think what they think because it is what they want to think (because it is motivated, because it is framed to appeal to their deeply held values, because of their cultural and ideological commitments), it seems there is nothing to do but continue the ideological wars, hoping for long-term victory through education (more manipulation?) and attrition from human mortality—a bleak picture indeed. Nor do we want to return to the days of deficit model thinking. Finally, we don't want to hide what scientific debate there is (in the hopes that doing so will bolster scientific authority), as open debate is part of what makes science robust and reliable, and making it seem otherwise is one of the flaws of the deficit model. We need another way forward.

Recognizing Legitimate Values in Science

There is a way out of this conundrum, a path that can be discerned by applying a lesson from the first lecture on science and values to science communication. The key lesson is that an influence of values on scientific reasoning need not be irrational (in the sense of being publicly defensible and based on public reasons). The trouble with the theories of the social scientists is that they presume a value-free ideal for science, that scientific reasoning should not be influenced by values, and that there is a clear dichotomy between accuracy-directed reasoning and motivated reasoning (to put it in Kunda's terms). Inductive risk considerations teach us that this is not the case, that pathways for motivated reasoning can be publicly defensible and deemed perfectly rational—and not just in the sense of pursuing one's self-interest. Further, we can tell whether or not values are playing a legitimate role by engaging with people and asking key questions.

When examining motivated reasoning through the lens of inductive risk, there are two possible interpretations—one that is problematic and one that is completely rational. The first interpretation is that people are ignoring evidence that goes *against* what they want to believe and emphasizing evidence that *fits* with what they want to believe (or already believe). Within the role-restriction ideal for values in science,[22] this employs the unacceptable direct role for values in reasoning about evidence, that the values ensure that you get to the desired (even if incorrect) conclusion. The second interpretation is that people are more worried about some risks of error than others and are adjusting their standards of proof accordingly. This interpretation utilizes inductive risk in science and emphasizes the indirect role for values in science, that values can and should help decide how much evidence is sufficient. If such values can create divergent views among experts, we should not be surprised that they can also create divergent views among the public. Further, the extent to which values are influencing the standards of proof, we can understand such reasoning as a rational use of values in evidential evaluation (for both experts and nonexperts).

The empirically based literature does not differentiate between these two ways in which social and ethical values can influence reasoning about evidence. So I can't tell which mode of reasoning people are actually employing when they are reasoning about science. I suspect it is a mixture of both. Sometimes, people are using unacceptable (biased or irrational) forms of motivated reasoning and, sometimes, people are using acceptable (rational) concerns arising from inductive risk. This dual possibility provides more than competing plausible explanations for the motivated reasoning phenomena. It also provides ways to move debate forward, beyond the impasse of ideological stalemate.

The lens of inductive risk clarifies a way to proceed with entangled political, ideological, and scientific debates, a way to engage in a dialogue with those who disagree with us. Let us take an example of two parties, A and B, who dispute the scientific support for Claim C. Both parties want to use the language of science to bolster their arguments, i.e., they both want to be seen as taking science seriously as a basis for their position. Party A argues that the science supports Claim C and we should proceed with policy on that basis, whereas party B argues that the science does *not* provide adequate support for Claim C (or argues that the science is indeterminate between Claim C and Claim F, or really supports F). We should all be familiar with such cases. We could be talking about anthropogenic climate change, vaccine safety, GMO safety, dietary recommendations, drug safety and efficacy, or evolution vs. creationism, among other contemporary debates.

The way these debates go is usually less than productive. Sometimes disputes turn into debates about specific evidential claims, with arguments and counter-arguments about methodological adequacy of particular studies. This can sometimes be productive, but often just turns into entrenched sides. More often, disputes turn into *ad hominem* attacks on who has received funding from whom, or who is biased by some influence, or who is being irrational, or who is really anti-science. When arguments are made on this basis, it is little wonder that they devolve into mutually loathing camps.

Inductive risk provides a different way to proceed. We can start with the presumption that everyone is acting rationally and is taking the evidence seriously. We can ask all parties (even our own—this is entirely reflexive) why they view the evidence as supporting their view. This should allow them to elucidate both the evidential and methodological basis for their view and the value basis, including why they view the evidence as sufficient. We can then ask all parties what evidence would change their minds. If what is at issue are value-based disagreements about evidential sufficiency (i.e., a proper indirect role for values regarding evidential sufficiency rather than an improper direct role), some hypothetical evidence should tip the balance to their opponents' position, thus illuminating the sufficiency standards others are using.

There are several discursive paths to follow at this point, depending on what the disputing parties say. For example, suppose party B says no evidence would ever change their mind. Then we can legitimately point out that their view is not scientifically based at all, that science has little or nothing to do with their stance, and that an ideological or value commitment external to science is what is driving their view. The debate can then properly focus on that commitment, what reasons we can have for or against it, and science can be left out of it. Imagine how much better the creationism debate would be if we could do this.

Suppose, on the other hand, that both parties can articulate hypothetical evidence that would change their mind. There are two directions to go with this information. One is to debate more carefully the values involved in the evidential sufficiency assessments. Why does party A have such lower standards than party B, or why is party B so demanding? Is one of the parties ignoring important implications of wrongly accepting or rejecting the claim? Are there some important methodological issues at stake?

Further, one can try to construct alternative policies that do not threaten the values at stake for one of the parties. For example, suppose we are talking about climate change and party B is worried about prematurely accepting anthropogenic climate change because of worries about a loss of personal freedom or a threat to private businesses. Instead of dismissing such concerns, we could work to ensure that shifting away from fossil fuels enhances individual freedom (e.g., through energy independence and decentralized energy generation), and that the business opportunities represented by such a shift are strongly supported by government policies through free-market mechanisms.

Finally, one could also attempt to collect the evidence that either party said they would find convincing. Perhaps with such evidence in hand, the dispute could be resolved, and people *would* change their minds.[23] Alternatively, the party that said it would be convinced could recant (or level additional methodological critiques). But doing so would require additional reasons targeted at the recent evidence, other than a distaste for the evidence. A failure to offer solid reasons for remaining skeptical, and for being able to say again what evidence would be convincing (which would still be required), or a repeated pattern for setting bars and then being unconvinced when the bars are cleared, should make us skeptical our interlocutors are engaging in the dispute with integrity. Such failures are evidence that their values are playing an improper direct role and should damage the credibility of those who exhibit such patterns. In short, depending on how the debate goes, lots of potential fruitful paths open.

By having a theory of how values can legitimately influence evidential assessment, we can provide a constructive path for engaging with people who disagree about evidential assessment. We don't have to just point out that they are probably cherry-picking data or agreeing with others who share their cultural identity, observations both of which are pretty insulting to a person's sense of intellectual integrity. We can use a theory of how values legitimately influence science to structure a respectful and (possibly) productive engagement.

And this is just one way the philosophical understanding of values in science can help with apparently ideological debates over science. Sometimes disputes rest on the kinds of studies are done or not done, which influences the values that shape the research agenda.[24] Acknowledging that there are gaps in the literature that need to be filled is important when addressing concerns about scientific claims. Openly discussing values in science provides avenues for moving the discussion forward, which is preferable to our current stalemates. We can do so by openly acknowledging the values in science, and the values at stake in science-based disputes, without disparaging the public or losing scientific integrity.

Scientists, rather than being frustrated by the divergent values of the public that make them resistant to scientific claims, should view those divergent values as a resource. It is those values that can redirect research efforts to problems the public thinks important or to gather evidence crucial for settling disputes by meeting everyone's evidential sufficiency standards. But having this kind of productive dialogue requires a public that understands scientific practice and the proper role for values therein. This will require both a shift in what science literacy should mean and recognized spaces for this kind of dialogue.

Rethinking Science Literacy

In order to generate the productive debate described in the previous section, note what is and is not needed for the public's baseline understanding of science. What is *not* needed is a passing grade on fact-based literacy tests. What *is* needed is an understanding of the process and practice of science. Indeed, we might rethink the name we give to this understanding. Rather than "literacy" (the ability to read and write), perhaps the skill we want for citizens with respect to science is competency or fluency, terms more reflective of an ability to engage appropriately and productively with science.

The most important thing for the public to understand about science is not a set of scientific facts, but its nature as an empirical, inductive, and critical process. Science is empirical because of the central importance of evidence gathered from interacting with the world (both social and natural). Scientists focus on gathering evidence to test the theories they develop. But what comes with this emphasis on the empirical is the inductive uncertainty of scientific knowledge. Any given claim[25] can be challenged (and refined or overturned) by future evidence. The evidence is never complete for any general scientific claim. It is the possibility for future evidential challenge that makes science so exciting for scientists—genuine discovery and novel ways of thinking about the world are always a possibility. And because science is empirical and inductive, it also must maintain a culture of critical interactions among science peers. Scientists have to be willing to challenge each other's work, to overturn longstanding views (if the evidence is there to do so), and to hold no claim above the critical fray. It is this critical culture, the social culture of science, combined with (and arising from) its evidential and inductive basis, that gives science its underlying epistemic authority.

The public must have a clear understanding that this is what science is. Teaching science as—and structuring literacy tests around—a fixed set of facts creates the opposite understanding. That science is an ongoing practice of investigation should be the first thing the public learns about science in grade school, rather than an understanding of science they encounter only once they reach graduate studies (for those who pursue them).

If the public embraced such an understanding of science, it would be less perplexing when experts disagree or when experts change their minds in the face of new evidence. Currently, members of the public express dismay or frustration with the instability of expert knowledge when confronted with experts in practice. But if the public understood that this was normal, that, in fact, this was science working properly, such frustration would likely be mitigated, if not resolved entirely.

If we focused on such an understanding of science, we could shift our science literacy tests accordingly. Doing so does not mean emphasizing only the qualitative open-ended questions of Miller's tests (discussed above) and losing the ability to have readily assessable and comparable measures. Recently, some social scientists have begun developing more appropriate scientific literacy tests, which require an understanding of scientific reasoning rather than scientific facts.[26] In these tests, one's understanding of how to control for causal confounders, how to be aware of problems such as drawing results from small numbers, and how, in general, to reason about scientific evidence is tested in easily assessable true/false questions. Such a test eschews fact-based science literacy testing and replaces it with the requisite understanding of the challenge of doing science and of reasoning about evidence. We do not have to give up readily comparable measures of scientific literacy when we shift to focusing on the process and reasoning of science.

In addition to having a more durable and appropriate goal for science literacy, there are potentially practical benefits to centering our literacy efforts on this kind of test. In developing their measures for scientific literacy, Drummond and Fischhoff found that those who do well on such tests do tend to accept the scientific consensus on contentious issues such as climate change more than those who do poorly. I do not mean to suggest that getting the public to agree with scientists is the gold standard for measures of scientific literacy. As noted above, there can be good value-based reasons for disputing scientific results as well. But we should take some comfort in the idea that those who understand better how scientists reason also tend to accept and agree with what scientists are trying to tell us.

Finally, having a public that understands the scientific process would enable us to refocus science education on how research gets done (including a country's science policy culture) and how that research gets used to make policy. Science civics should be a core part of every citizen's education, including how values shape what research gets conducted as well as assessments of evidential sufficiency. That values legitimately shape scientific practice in these ways should be part of the basic education regarding science. The public may, in the process, gain tools to critically examine current scientific practices, to query the direction and institutional structures that shape research agendas, and to question the science advice apparatus more thoroughly. Rather than view this as trouble, this is what we should hope for: a scientifically informed and democratically engaged public. The question then becomes, how do we structure interactions between the public and the scientific community in practice?

The Requisite Social Structures and Practices

One big institutional shift will need to be in K-12 education. Rather than focusing on training possible future scientists and getting core disciplinary facts across to students, primary and secondary education needs to be focused on what science is (as described above), how values shape science, and how policy structures interact with science. Science civics becomes a central rather than peripheral topic. There are already reform efforts underway, which focus on guiding students through actual discovery processes rather than fact-based learning (they learn the facts they need along the way in the discovery process). Efforts like the AAAS's Project 2061[27] try to emphasize a more process-based approach to science education. But such efforts are hampered by the tremendous institutional inertia of the educational system, by continued insistence on disciplinary fact-based learning as well as standardized testing based on the current inadequate system. Changing all this will be challenging, but we should be heartened by the fact that kids really love doing science and have a tremendous capacity for it. My favorite exemplar of what is possible is the Blackawton Bee study, where second-graders (yes, grade 2!!) performed and subsequently published a study on bee behavior. Their conclusion: "Science is cool and fun because you get to do stuff that no one has ever done before."[28] Note that this was not some handpicked cadre of exceptional students, but a regular class at a regular elementary school. Every kid can *do* science if given the chance. It is far more important that kids get the chance to experience scientific inquiry than to master scientific facts and our educational system should reflect this.

But we cannot wait for the change in the educational system and for future generations to pass through it before engaging democratic publics better in science and scientific issues. There are a range of techniques and structures that are currently being employed, and from these we can draw, test, and expand the possibilities.

Citizens are already engaged in "citizen science" projects from bioblitzes, to astronomical observation and backyard data collection (bird counts, ice rink records, weather stations) to online engagement (through games like FoldIt or video scanning from undersea cameras). These are of great assistance to scientists, but they do not engage citizens in the judgments needed to set up a scientific study or to assess evidence. Instead, they use citizens as data collection mechanisms. Perhaps such projects can be opened to include citizen input on the direction of future studies, thus drawing citizens deeper into the scientific process.

A deeper kind of engagement with scientific issues can be seen in a range of deliberative forums deployed over the past two decades. Citizen consensus conferences, Fishkin's deliberative polls, and deliberative forums like the World-

WideViews project[29] bring groups of citizens (often demographically representative) together to discuss a particular science-based policy issue, informed by experts.[30] Sometimes citizens come to consensus and sometimes they vote on what to do, and thus divergent views are preserved in the record. Such conferences generally have produced sensible and well-reasoned assessments of issues (even if they have had uneven success in their influence on policy). Citizens in such spaces become both better informed about a particular issue and more reflective about how they would like their community or society to deal with it. (They tend to think and argue more like citizens than consumers, to draw from Mark Sagoff's classic distinction.[31]) As such, they raise the level of democratic discourse about a particular issue while in general raising the level of democratic engagement among citizens. These are good things.

While such forums are useful for assessing the available research and to improve the discussion of such issues in democratic societies, they tend not to influence future research or how experts think about their work. We can have even more robust engagement between citizens and experts through collaborative research projects. In these kinds of processes, citizens (often stakeholders as opposed to representative samples) work with experts to produce the knowledge needed to address a particular issue. It is important that such processes be open and transparent, that a range of stakeholders be involved from the start, and that trust be built and maintained between citizens and experts over time. When this works (and it has worked well in the past[32]), value judgments and evidence both get aired and examined, different evidential standards can be met (so that everyone is satisfied with the final result), and expertise can soundly and accountably inform policy. There is robust learning (from two-way dialogue) for all parties involved. This is also near-ideal public interest science,[33] and thus needs to be better supported by funding mechanisms. But this involves getting experts repeatedly in the room with engaged citizens and stakeholders, and thus can be temporally and spatially constraining. Although such a process may not be applicable to every issue, when scientific expertise is relevant and needed for local issues, this is a near ideal way to approach the problem of public involvement.

Finally, there are ongoing experiments with other forms of public-expert interactions. Colleagues, for example, are experimenting with material deliberation, with ways in which citizens can interact with the stuff of science, and thus think through what kinds of possibilities and directions new technologies can take.[34] There are also efforts to embed social scientists and humanists in science labs so they can work with scientists to help them understand the potential implications of—and the nascent value judgments in—their work, and thus create the potential to bring such work more in line with the needs and interests of the public.[35] In the field of responsible innovation, such experiments are helping to uncover what works, when, and why. In general, the recognition of the importance of social and ethical values in shaping what science gets done should provide further impetus to "upstream engagement" techniques in the responsible innovation sphere.

These kinds of structures can produce precisely the kind of dialogue—the kind of two-way communication—that is needed for science communication to be successful. Science is not a value-neutral, universally authoritative fact-producing machine in these processes. Instead, citizens and scientists work together to decide which science should be done and how it should proceed, and when a study is done (i.e., when the evidence is enough). As such, the science and expertise are inherently made democratically accountable.

As these kinds of mechanisms for interactions become a normal part of democratic governance, we can expect to see a concomitant shift in the more traditional forms of science communication, such as science journalism and science museums. Journalists would have more to cover than embargoed scientific results in scientific journals, with reports coming out of processes like citizen juries and collaborative analyses. Science journalism is already shifting away from the simplistic view that objective reporting means reporting two opposing views. More appropriate would be reporting on reasons why experts agree or disagree with a finding, and the value commitments involved in these assessments. (This could also make science reporting even more interesting to the public!) And science museums are already experimenting with more interactive science communication methods, for example, with respect to nanotechnology.[36] Our "informal" science education system can readily adapt to more dialogue and value-transparent modes of science communication. Further, social media platforms can be used to support dialogue across scales where meeting face-to-face is not feasible.

So, there are a multitude of practical ways to change how we conduct science communication to include a two-way, interactive component. However, pursuing these avenues also means changing how we think of science in public debate.

What This Means for the Public Role of Science

I have been describing a lot of promising avenues, both conceptual and practical, for science communication post-value-free ideal and post-deficit model. But there is one thing that must be relinquished regarding how we think about science in public discourse, and I suspect that is a source of great anxiety for many.

What we must give up is the idea that science is a value-neutral source of authoritative statements. If values are important in shaping the research agenda of science and important in deciding when evidence is sufficient, then only scientific claims informed by acceptable values (and the definition of acceptable will change depending on whom you are talking to in a pluralist society) will be, and *should* be, authoritative. The ideal of science having a universally authoritative voice, above the fray of our political differences, must be surrendered.[37]

Now, we know we haven't been living in a world where science has universal authority for some time. I do not think the public contestation of science is a new thing—I am skeptical there was ever a golden age when scientists were universally listened to, when their advice was always heeded, when expertise was unproblematic. Indeed, once we had specialized expertise, we had contested expertise (at least so it seems to me from the historical record). But the sense that we are now in an age of "post-normal" science, and that science is under siege as never before, is clearly pervasive. And it might seem that the understanding of science I have been advocating, of a value-saturated endeavor, where shared values are a *part* of the basis for accepting science, just gives in to these trends.

I think, however, that relinquishing the idea that science can act as a value-neutral part of our public discourse is no great loss when 1) we can have more productive debates about science without that idea, and 2) that idea does not seem to be helping in our practices currently. We might be losing the ideal of science as a source of universal public reason, but we are gaining an ideal of discourse about evidentially based issues in practice.

Science, under this vision of its role in society and its relationship to the public, is no longer a source of empirical claims with automatic authority. The authority of science rests on several bases simultaneously (as noted in the first lecture). It is based partly on the integrity of individual scientists (keeping values to their proper roles), partly on the communal critical practices of scientists, partly on the instrumental success of science in human endeavors, and partly on shared values. This means that scientists should expect some members of the public (whose values are threatened by their findings) to be hesitant, or even resistant, to accepting their findings. The right response is not to bemoan a lack of factual knowledge about science, but to engage those publics, to find out why they view the findings as problematic, to ascertain what evidence would convince them of the scientific work or which questions they want answered, to craft ways forward. We are starting to construct social spaces in which this can happen, and this is where future efforts should lie.

In short, science can retain its place as a general authority in society by acknowledging the complex basis of its authority. It is authoritative not because it is value-free or produces permanent facts, but because it is a critical, empirical endeavor conducted by a community structured to criticize itself, and because of values shared with the society in which its work is embedded. Specific claims are always open to critique, both within and outside of science, on both an empirical and valuational basis. How those criticisms unfurl depends on the kind of criticism. The responsiveness to that criticism is part of the basis for the authority of science. Displaying the resulting dialogue needs to be seen not as just part of science communication, but part of the democratic discourse essential to our science- and technology-laden societies.

Conclusion

In a sense, the scientists are right: the public's lack of knowledge about science is a major problem. But it is not a lack of factual knowledge, of discrete and easily measured bits of information that is the key deficit. It is rather a lack of understanding of what science is, as an endeavor, that is the problem. Scientific literacy tests are geared toward the wrong objective, and as such, undermine the public's ability to engage with science. We can remake our sense of what civic science literacy should be, and what our educational system should look like, accordingly.

Further, shared values are a key and legitimate resource for the authority of particular scientific claims. The public needs to believe that scientists care about what they *should* care about, that they are asking the right questions, and that they are appropriately assessing the evidence in order for science to be authoritative on particular issues. We can craft practices for conducting science communication in more engaged, deliberative ways, but we need to have a clear idea of how values are legitimately part of science, and how they can serve as both a source of contestation and acceptance of scientific claims, in order for those practices and the institutions that support them to be properly structured.

Science communication should not be from only one perspective or one-directional. Publics in democratic societies are plural, heterogeneous. While it may be challenging to honestly engage such publics, I hope I have provided a sense of how to do so here without being either deceptive or manipulative. Whether proceeding with these practices helps build trust remains to be seen. But whenever frustration looms, the scientific community should recall that the heterogeneity of the public is also a resource for science—for doing science *better*. For it is from this heterogeneity that science draws the diverse voices, the different ways of looking at a problem or thinking to look where no one has looked before that makes science better.

Indeed, the differences between us, the gaps in values among citizens, the contestation in politics, the spaces between evidence and theory in science—all of these gaps that frustrate the urge to create a smooth and perfect foundation, the one perfect communication, the finished and complete political system—need to be celebrated as resources for legitimacy. Rather than try to fill or eliminate the gaps, we should see them as the space that provides the moment for criticism, for new ideas, for disagreement that produces better understanding down the road. Even the gap between descriptive and normative statements is important for being able to conceive of a world better than the one we are currently in. Our world is riddled with gaps, and it is through our willingness to look across them, to engage across them, that we can make a robust, if unsettled, future.

Notes

[1] See Keay Davidson, *Carl Sagan: A Life* (New York, NY: Wiley, 1999), and Karen Jane Sorensen, "Carl Sagan's *Cosmos*: The Rhetorical Construction of Popular Science Mythology," Dissertation, North Dakota State University (April 2013).

[2] Pew Research Center, "Public and Scientists' Views on Science and Society" (2015), available online: https://www.pewresearch.org/internet/wp-content/uploads/sites/9/2015/01/PI_ScienceandSociety_Report_012915.pdf.

[3] Robert C. Davis, *The Public Impact of Science in the Mass Media: A Report on a Nation-Wide Survey for the National Association of Science Writers* (Ann Arbor, MI: Institute for Social Research, University of Michigan, 1958).

[4] Ibid., 135.

[5] Matthew C. Nisbet and Dietram A. Scheufele, "What's Next for Science Communication? Promising Directions and Lingering Distractions," *Botany* 96, no. 10 (2009): 1767–1778.

[6] Jon D. Miller, "The Measurement of Civic Scientific Literacy," *Public Understanding of Science* 7, no. 3 (1998): 204.

[7] Quoted in ibid., 204.

[8] G. Wayne Clough, *Increasing Scientific Literacy: A Shared Responsibility* (Washington, DC: Smithsonian Institution, 2011), 6.

[9] For example, see the table produced by the National Science Foundation, http://www.nsf.gov/statistics/seind14/index.cfm/chapter-7/tt07-08.htm

[10] Jon D. Miller, "The Measurement of Civic Scientific Literacy," *Public Understanding of Science* 7, no. 3 (1998): 205.

[11] Matthew C. Nisbet and Dietram A. Scheufele, "What's Next for Science Communication? Promising Directions and Lingering Distractions," *Botany* 96, no. 10 (2009): 1767–1778.

[12] See also Brian Wynne, "May the Sheep Safely Graze? A Reflexive View of the Expert-Lay Knowledge Divide," in *Risk, Environment and Modernity: Towards a New Ecology*, Scott Lash, Bronislaw Szerszynski, and Brian Wynne, eds. (London, UK: SAGE Publications, 1996), 44.

[13] Toby Bolsen, James N. Druckman, and Fay Lomax Cook, "Citizens', Scientists', and Policy Advisors' Beliefs about Global Warming," *The ANNALS of the American Academy of Political and Social Science* 658, no. 1 (2015): 271–295.; Dan M. Kahan, Ellen Peters, Maggie Wittlin, Paul Slovic, Lisa Larrimore Ouellette, Donald Braman, and Gregory Mandel, "The Polarizing Impact of Science Literacy and Numeracy on Perceived Climate Change Risks," *Nature Climate Change* 2, no. 10 (2012): 732–735.

[14] This is one of the facts used to measure scientific literacy in Dan M. Kahan, Ellen Peters, Maggie Wittlin, Paul Slovic, Lisa Larrimore Ouellette, Donald Braman, and Gregory Mandel, "The Polarizing Impact of Science Literacy and Numeracy on Perceived Climate Change Risks," *Nature Climate Change* 2, no. 10 (2012): 732–735, for example.

[15] Here Kahan draws from the work of Mary Douglas and Aaron Wildavsky on the anthropology of risk. See Dan M. Kahan, Hank Jenkins-Smith, and Donald Braman, "Cultural Cognition of Scientific Consensus," *Journal of Risk Research* 14, no. 2 (2011): 147–174.

[16] See, e.g., Matthew C. Nisbet and Chris Mooney, "Framing Science," *Science* 316, no. 5821 (2007): 56.

[17] Ziva Kunda, "The Case for Motivated Reasoning," *Psychological Bulletin* 108, no. 3 (1990): 483.

[18] Ibid., 488–489.

[19] Ibid., 490.

[20] Dan M. Kahan, Ellen Peters, Maggie Wittlin, Paul Slovic, Lisa Larrimore Ouellette, Donald Braman, and Gregory Mandel, "The Polarizing Impact of Science Literacy and Numeracy on Perceived Climate Change Risks," *Nature Climate Change* 2, no. 10 (2012): 733–734.

[21] Matthew C. Nisbet and Dietram A. Scheufele, "What's Next for Science Communication? Promising Directions and Lingering Distractions," *Botany* 96, no. 10 (2009): 1776.

[22] Discussed under ideal #2 in Lecture 1, this volume, 24–25.

[23] I recount an actual case of this in Heather Douglas, "Politics and Science: Untangling Values, Ideologies, and Reasons," *The ANNALS of the American Academy of Political and Social Science* 658, no. 1 (2015): 296–306.

[24] Discussed in greater depth in Lecture 2, this volume, 67–93.

[25] Including claims about the usefulness or reliability of particular methodologies.

[26] Caitlin Drummond and Baruch Fischhoff, "Development and Validation of the Scientific Reasoning Scale," *Behavioral Decision Making* 30, no. 1 (2017): 26–38. [First published online October 12, 2015.]

[27] See AAAS, "Project 2061," available online: https://www.aaas.org/programs/project-2061.

[28] P. S. Blackawton, S. Airzee, A. Allen, S. Baker, A. Berrow, C. Blair, M. Churchill, J. Coles, R. F.-J. Cumming, L. Fraquelli, C. Hackford, A. Hinton Mellor, M. Hutchcroft, B. Ireland, D. Jewsbury, A. Littlejohns, G. M. Littlejohns, M. Lotto, J. McKeown, A. O'Toole, H. Richards, L. Robbins-Davey, S. Roblyn, H. Rodwell-Lynn, D. Schenck, J. Springer, A. Wishy, T. Rodwell-Lynn, D. Strudwick, and R. B. Lotto, "Blackawton Bees," *Biology Letters* 7, no. 2 (2011): 168–172.

[29] See World Wide Views, available online: http://wwviews.org.

[30] Frank Fischer, *Democracy and Expertise: Reorienting Public Policy* (New York, NY: Oxford University Press, 2009), Chap. 3;

James S. Fishkin, *When the People Speak: Deliberative Democracy and Public Consultation* (New York, NY: Oxford University Press, 2009).

[31] Mark Sagoff, "At the Shrine of Our Lady of Fatima or Why Political Questions Are Not All Economic," *Arizona Law Review* 23 (1981): 1283–1298.

[32] See, e.g., George J. Busenberg, "Collaborative and Adversarial Analysis in Environmental Policy," *Policy Sciences* 32, no. 1 (1999): 1–11, and the Pew Initiative Report on Bt corn and the monarch butterfly, "Three Years Later: Genetically Engineered Corn and the Monarch Butterfly Controversy," available online: https://www.pewtrusts.org/~/media/legacy/uploadedfiles/wwwpewtrustsorg/reports/food_and_biotechnology/vfbiotechmonarchpdf.pdf.

[33] Argued for in Lecture 2, this volume, 67–93.

[34] Sarah R. Davies, Cynthia Selin, Gretchen Gano, and Ângela Guimarães Pereira, "Citizen Engagement and Urban Change: Three Case Studies of Material Deliberation," *Cities* 29, no. 6 (2012): 351–357.

[35] Erik Fisher and Daan Schuurbiers, "Socio-Technical Integration Research: Collaborative Inquiry at the Midstream of Research and Development," in *Early Engagement and New Technologies: Opening Up the Laboratory*, Neelke Doorn, Daan Schuurbiers, Ibo van de Poel, and Michael E. Gorman, eds. (Dordrecht, Netherlands: Springer, 2013), 97.

[36] Ai Lin Chun, "The Shifting Roles of Museums," *Nature Nanotechnology* 9 (2014): 724.

[37] See also Dan M. Kahan, "The Cognitively Illiberal State," *Stanford Law Review* 60 (2007): 115–154.

Commentary on Lecture 3

SCIENCE'S IMAGE: BRINGING DOUGLAS INTO FOCUS

Eric Schliesser

In these set of comments, I first introduce a bit of special jargon—what I call "an image of science"—in order to discuss some core commitments that inform Douglas' science communication proposal. Second, I highlight three big-picture commitments I discern in Heather Douglas' project: (1) inductive risk, (2) epistemic (so-called "Knightian") uncertainty, and (3) a Popperian (albeit not falsificationist!) sensibility. In recent times, interest in (1) has been revived by Douglas and so should not come as a surprise.[1] Third, I call attention to these three commitments in order to argue that in Douglas' science communication proposal—the topic of the third Descartes lecture—her image of science is a normative one, despite her tendency to present it as empirical. I argue that this entails that Douglas needs to develop her science communication proposal to cover situations that do not live up to her normative ideal and to make more explicit the significance of her normative commitments.

An Image of Science

A professional society of recent vintage, the Committee for Integrated HPS, which is dedicated to "the integration of the history of science and the philosophy of science," is organized around "the conviction that the common goal of understanding of science can be pursued by dual, interdependent means."[2] Even so, anyone familiar with developments in recent history of science and philosophy of science will sometimes wonder if the two disciplines talk about the same thing. In fact, when philosophers, historians, and sociologists of science, as well as, say, policy makers, speak of "science," it's not always clear what they have in mind. (And, of course, there are plenty of philosophers of science who think that science is plural in some sense.[3]) To illustrate this point, there is even a whole subdiscipline within the philosophy of science—the so-called "philosophy of scientific practice" (SPSP),[4] with its own distinct professional society—that takes itself to be correcting traditional philosophy of science, which, it is said, disregarded "scientific practice." The mission statement of SPSP suggests that "the concern with practice has always been somewhat outside the mainstream of English-language philosophy of science."[5]

Here, in order to engage with some of Douglas' policy proposals, I introduce some special vocabulary to discuss one distinct element of science when people talk of "science." An *image of science* is (a) a list of characteristics that function as a shorthand for representing various scientific activities and products (or scientific expertise), and (b) it is used in debates, where at least one side relies on the epistemic authority of science (in some sense). The image is often accompanied by both (c) a privileged list of scientific or epistemic virtues, and often relies on (d) lots of tacit and suppressed commitments about the nature of knowledge, the nature of reality, the nature of society, and the nature of science and scientific expertise.

While such images of science may be revealed in polemical contexts, they can also circulate and be taught in ordinary disciplinary training or, say, journalistic circumstances. They may be expressed in methodological debates within or about science, in social debates in which science plays some role or another, and within philosophical debates. I call it an "image" because while it generally presents itself as representing or characterizing science (or some sciences) as such, it nearly always abstracts away from other nontrivial features of science(s) that may well be pertinent.

Let me offer an example of such an image of science at work in Douglas' oeuvre. In her wonderful 2009 book, she claims that scientists are (a) "in near constant communication" and (b) "competition with other scientists;"[6] that (c) "ideas" spread "rapidly" and (d) "scientists" discuss "pitfalls" "readily."[7] As it happens, these elements taken together are characteristic of thinking of science as involving a kind of efficient market of ideas. Now, this is a useful idealization and image of science for some purposes.[8] But as a representation of reality, it ignores, for example, the significance of secret (i.e., classified) work and industry-sponsored science (often embargoed). In addition, this image ignores the incentives (and, perhaps, cultural issues) that generate now-familiar replication issues and confirmation bias in the literature.[9]

That science is or is not always akin to an efficient market of ideas can be more or less important depending on the use of the image. Versions of this image sometimes get used to argue for deference to the scientific community, which is presented as a self-correcting enterprise; some-times the self-correcting image is introduced in order to contain criticism.[10] The image of science in Douglas' 2016 Descartes Lectures does not rely on an efficient market of ideas, for Douglas provides examples of scientists that do not self-correct and fail to challenge each other. This makes one wonder whether she has changed her mind between the book and the Descartes Lectures, or whether we should understand her as offering an ideal type (in the Weberian sense) in one analysis and a kind of second best in the more recent Lectures.

A key passage for our present purposes in the third Descartes Lecture is this:

> *The public must have a clear understanding that this is what science is. Teaching science as – and structuring literacy tests around – a fixed set of facts creates the opposite understanding. That science is an ongoing practice of investigation should be the first thing the public learns about science in grade school, rather than an understanding of science they encounter only once they reach graduate studies (for those who pursue them).*[11]

In the passage, Douglas offers two (partial) images of science: the first one (to be rejected) is focused on a fixed set of facts; the second, more preferable one, is committed to science being an ongoing practice of investigation. The two images are not necessarily in logical conflict with each other although they are clear alternatives. (They become contradictory if one assumes that the second image is also committed to a rather thoroughgoing fallibilism about facts, but that's not a standard move.)[12]

In order to forestall misunderstanding, one may wonder that if the two images are not contradictory, why call them alternatives? I do so not just to do justice to Douglas' (correct!) use of "opposite," but also because an image of science presents science as being a certain way and leaves no presentational (and psychological) room for the other ways science may be.

Douglas' proposal is, in fact, to train not just a (small number) of would-be scientists, but to teach the "nature of science" (and the nature of "scientific reasoning") to the public at large. By Douglas' lights, the second image captures essential characteristics of science that the first does not. So Douglas' proposal consists of both a suggestion about content (that is, her image of science) and audience size. The payoff of Douglas' proposal is that the right image of science would produce better public conversations and deliberations about science as well as better science.[13] In fact, Douglas relies on the idea that her image of science is already familiar to those with advanced training in the sciences—"graduate studies (for those who pursue them)." In the old image of science (the one Douglas rejects), science is an uncontested (and neutral) source of facts and knowledge; in her proposed, alternative image of science, it becomes a collaborative, open-ended enterprise with the public.[14]

In Douglas' hands, the proposal also requires that within the intellectual division of labor, folks like us—philosophers of science—end up playing a key role in mediating between scientists, who are busy investigating stuff, and the public. I am not convinced most philosophers are equipped to engage with the public,[15] although Douglas proves an admirable exception to the rule. Before I develop more fully a critical response to Douglas, it may be useful to highlight some other fundamental characteristics of her image of science.

Three Big-Picture Commitments

One reason to welcome Douglas' Descartes Lectures is that it is clear that Douglas' image of science does not just rely on her arguments about (1) inductive risk, but also embraces (2) epistemic uncertainty, and (3) a Popperian (albeit not falsificationist!) sensibility. Douglas' work on inductive risk is subtle and too complex to review here. I just note that she has shown the many ways values permeate science and, simultaneously, do not undermine scientific authority, and that seemingly irrelevant social consequences can matter greatly in our evaluation(s) of science. Here I focus on (2) and (3).

The package of commitments is revealed in a passage where Douglas sums up what image of science "the public" must understand about science:

> *The most important thing for the public to understand about science is not a set of scientific facts, but its nature as an empirical, inductive, and critical process.* (A) *Science is empirical because of the central importance of evidence gathered from interacting with the world (both social and natural).* (B) *Scientists focus on gathering evidence to test the theories they develop. But what comes with this emphasis on the empirical is,* (C), *the inductive uncertainty of scientific knowledge.* (D) *Any given claim can be challenged (and refined or overturned) by future evidence.* (C*) *The evidence is never complete for any general scientific claim. It is the possibility for future evidential challenge that makes science so exciting for scientists – genuine discovery and novel ways of thinking about the world are always a possibility. And because science is empirical and inductive,* (E), *it also must maintain a culture of critical interactions among science peers.* (E*) *Scientists have to be willing to challenge each other's work, to overturn longstanding views (if the evidence is there to do so), and,* (D), *to hold no claim above the critical fray.* (F) *It is this critical culture, the social culture of science, combined with (and arising from) its evidential and inductive basis, that gives science its underlying epistemic authority.*[16]

Douglas' commitment to epistemic (or "inductive") uncertainty within science is revealed by two commitments: first her rather thoroughgoing fallibilism – science is never fully secure from possible revision. In addition, on her view, future science can be very surprising.[17] It can overturn what we take for granted about reality (e.g., C, C*, D). Such ideas about "inductive uncertainty" have a long pedigree in philosophy, including areas of philosophy very friendly to the authority of science (e.g., Russell).[18] I tend to associate them with so-called Knightian uncertainty.[19]

An embrace of Knightian uncertainty within scientific practice generates sources of "excitement" and a joy in "novelty," as Douglas emphasizes, but it also tends to promote a stance of epistemic humility and receptivity toward nature and others. It helps guard against expert overconfidence. Perhaps it's because her image of science quite obviously promotes such virtues that Douglas thinks this image of science would produce better public conversations and deliberations about science, for this image prevents scientists from being mere expositors or preachers—they can engage the public with a spirit of open-mindedness that allows them to learn, in principle, from any encounter.

In fact, Douglas' image of science does not merely embrace some characteristic virtues for individual scientists, but she is also explicit that it must be accompanied by a particular "culture," one that embraces (E), "critical interactions among scientists." I tend to associate this (D and E) with a Popperian mindset.[20] To simplify, in the Popperian image of science one does not focus on confirming one's scientific concepts or theories, but in stress-testing them.[21] The Popperian image of science emphasizes looking for evidence that can change one's mind and promote discussion, even critical disagreement.[22] Popper's image has come to be associated with falsificationism, but from the present perspective, that's really just a symptom or symbol of a commitment to a critical culture. The key to the Popperian image is not falsificationism, but rather something akin to (D), "any given claim can be challenged (and refined or overturned) by future evidence."

Before one assumes that Douglas' image of science is uncontroversial (and, of course, by associating it with Popper I imply it is not beyond controversy), it is worth being reminded of an alternative image of science that has been far more influential within scientific culture and discussions about it (albeit rarely so explicitly stated):

> *An extremist is an intellectual lunatic – allowed loose if he does not communicate violence, but without an admission ticket to ordinary discourse. There is merit in excluding the lunatic from the discourse. Occasionally the lone dissenter with the absurd view will prove to be right – Galileo with a better scheme of the universe, a Babbage with a workable computer – but if we gave each lunatic a full, meticulous hearing, we should be wasting vast time and effort.*[23]

Stigler won a Nobel Prize in economics (1982); he was himself a keen historian and sociologist of science. This image of science treats science as a consensus generating device. It's a familiar image because it is associated with Kuhnian philosophy of science.[24] Recall from Kuhn's *Structure,*

> *To a very great extent the "term" science is reserved for fields that do progress in obvious ways. Nowhere does this show more clearly than in the recurrent debates about where one or another of the contemporary social sciences is really science...they will cease to be source of concern not when a definition is found, when the groups that now doubt their own status achieve consensus about their past and present accomplishments. It may, for example, be significant that economists argue less about whether their field is a science than do practitioners of some other fields of social science.*[25]

In Stigler's passage we can discern a cost-benefit argument: time and attention is scarce, so it is pointless to listen to everybody. This implies that not all objections will be heard by the scientific community. Even if one stipulates that most possible objections are silly, for this image it's okay to ignore some excellent objections! This image of science treats (potentially revolutionary) dissenters as outside the peer group—it leaves little room for heterodox views.

The foreseeable danger of this image of science is that science can generate a strong status quo bias because potentially fundamental objections are not explored.

By contrast, in Douglas' image of science the (F) "authority" of science does not rest on it being a consensus-generating device, but rather rests on its (1) critical culture (D and E) and ultimately the (2) evidential basis of claims (testing) (viz., A and B). That is, according to Douglas, when science is challenged in public or when we need a proper source to help us figure out what is to be done, we should be armed with this image of science. But by calling it "'Popperian," I have also indicated that it is not a self-evident image of science.

Heather Douglas' Image of Science Reconsidered

I understand Douglas' image of science as an attractive, normative ideal, one that should be taught, as a normative ideal, not just to the public, but also to the scientists, grant agencies, and policy makers, so that they (the public, scientists, funding agencies, and politicians) can embrace and promote this image of science as much as possible.

It should be clear from the tenor of my remarks that I do not believe that Douglas' image of science is an accurate representation of reality. To be sure, Douglas' image of science draws on themes and norms immanent in scientific practice(s). But it is somewhat flattering to this practice. There are incentives and norms that prevent the full realization of science being a "critical culture."[26]

Constraints on space prevent me from offering full details of empirical sociology of knowledge here that would substantiate the suggestion in the previous paragraph, but it is worth reminding ourselves of the dual role of (refereed journal) publication in scientific culture: to simplify, it is both a public record of the status of certain scientific claims as well as a means on the path to jobs, promotion, grants, and status. Because the credit economy in which scientists operate can sometimes be a zero-sum game, it is not surprising that the (anonymous) referee and gatekeeping process does not always encourage a critical culture. I am not suggesting that this is the most important bottleneck against the development of a more vibrant critical culture in science—I suspect that the outsized influence of lab directors over the academic careers of their junior staff is more important. All I am implying is that we should not idealize the scientific status quo.[27]

In fact, I would urge that an endorsement of Douglas' image of science would encourage us to facilitate the development and adherence of norms and incentives that produce more "critical interaction" among scientists. And even when there is critical interaction among scientists, we also need to create incentives that prevent closing of ranks when science is challenged from without.

This much Douglas should be able to easily accommodate. Even if Douglas thinks that scientific reality is closer to the normative ideal that her image of science presupposes (than I think is the case), I am calling attention to the fact that we need a proposal to cover situations that do not live up to her normative ideals. Because human nature is the way it is, one can easily imagine that Douglas' image of science will be abused in polemical contexts; with funding and authority on the line, there will be a recurring temptation to stifle public criticism by suggesting that existing science has a critical culture and that, in principle, any given claim can be challenged. Because nonexperts are, by definition, in a bad position to evaluate such claims about the nature of scientific practice, we need mechanisms (norms and incentives) by which we can promote and check that existing science has a critical culture and that any given claim can be challenged. Perhaps Douglas' image of science can shape better practice.

Notes

[1] See, e.g., Heather E. Douglas, "Inductive Risk and Values in Science." *Philosophy of Science* 67, no. 4 (2000): 559–579, and Heather Douglas, *Science, Policy, and the Value-Free Ideal* (Pittsburgh, PA: University of Pittsburgh Press, 2009).

[2] Committee for Integrated HPS, "Our mission statement on integrated HPS," available online: http://integratedhps.org/about.html.

[3] See, e.g., Patrick Suppes, "The Plurality of Science," *PSA: Proceedings of the Biennial Meeting of the Philosophy of Science Association* 1978 (1978): 3–16.

[4] Joseph Rouse, "Kuhn's Philosophy of Scientific Practice," in *Thomas Kuhn*, Thomas Nickles, ed. (New York, NY: Cambridge University Press, 2003), 101.

[5] See Society for Philosophy of Science in Practice, "Mission statement," available online: http://philosophy-science-practice.org/en/mission-statement/. As it happens, Heather Douglas and I are friendly, fellow travelers of both Committee for Integrated HPS and SPSP (as well as the PSA, EPSA, and HOPOS to mention a few other groups among the alphabet soup of our affiliations).

[6] Heather Douglas, *Science, Policy, and the Value-Free Ideal* (Pittsburgh, PA: University of Pittsburgh Press, 2009), 93.

[7] Ibid., 83.

[8] For critical discussion when it comes to moral responsibility of scientists and scientific communities, see Merel Lefevere and Eric Schliesser, "Private Epistemic Virtue, Public Vices: Moral Responsibility in the Policy Sciences," in *Experts and Consensus in Social Science*, Carlo Martini and Marcel Boumans, eds. (New York, NY: Springer, 2014), 275.

[9] Neal S. Young, John P. A. Ioannidis, and Omar Al-Ubaydli, "Why Current Publication Practices May Distort Science," *PLoS Med* 5, no. 10 (2008): e201.

[10] Cf. Art Markman, "Why science is self-correcting," *Psychology Today* (Aug. 10, 2010), available online: https://www.psychologytoday.com/blog/ulterior-motives/201008/why-science-is-self-correcting with John P. A. Ioannidis, "Why Science Is Not Necessarily Self-Correcting," *Perspectives on Psychological Science* 7, no. 6 (2012): 645–654.

[11] Heather Douglas, this volume, 138, emphases added.

[12] Below I present evidence, where Douglas seems to embrace such thoroughgoing fallibilism about facts, although I think she is best understood as defending thoroughgoing fallibilism about what is *claimed* about facts.

[13] It is possible for a false image of science to produce better science, but that's not being offered here.

[14] Obviously, one can agree with Douglas that science is best understood as an ongoing practice of investigation, and also think (in opposition to her views) that it is an elite enterprise to which the public has little to contribute. But that's not my present position.

[15] Eric Schliesser, "Nothing Much for Philosophers," *Philosophy of the Social Sciences* 48, no. 1 (2018): 40–46.

[16] Heather Douglas, this volume, 137–138, capital letters added to facilitate discussion—ES.

[17] See, e.g., Hans-Jorg Rheinberger, *Toward a History of Epistemic Things: Synthesizing Proteins in the Test Tube* (Stanford, CA: Stanford University Press, 1997).

[18] Paul Hager, "Russell's Method of Analysis," in *The Cambridge Companion to Bertrand Russell*, Nicholas Griffin, ed. (New York, NY: Cambridge University Press, 2003), 310.

[19] Frank H. Knight, *Risk, Uncertainty, and Profit* (Boston, MA: Houghton Mifflin, 1921).

[20] See also Heather Douglas, *Science, Policy, and the Value-Free Ideal* (Pittsburgh, PA: University of Pittsburgh Press, 2009), 153.

[21] The version of Popperianism I am presenting here is much indebted to Abraham D. Stone, "On Scientific Method as a

Method for Testing the Legitimacy of Concepts," *Review of Contemporary Philosophy* 8 (2009): 13–48.

[22] Popper himself recognized that much of what Kuhn calls "'normal"' science does not fit his ideal. See Karl R. Popper, "Normal Science and its Dangers," in *Criticism and the Growth of Knowledge: Proceedings of the International Colloquium in the Philosophy of Science, London, 1965, Volume 4,* Imre Lakatos and Alan Musgrave, eds. (New York, NY: Cambridge University Press, 1970), 51.

[23] George J. Stigler, "The Unjoined Debate," in *The Citizen and The State: Essays on Regulation,* (Chicago, IL: University of Chicago Press, 1975), 3–4.

[24] I have recounted Stigler's interest in, and partial anticipation of, Kuhn elsewhere. For the context and significance of this passage, see Eric Schliesser, "Inventing Paradigms, Monopoly, Methodology, and Mythology at 'Chicago': Nutter, Stigler, and Milton Friedman," *Studies in History and Philosophy of Science Part A* 43, no. 1 (2012): 160–171, and Eric Schliesser, "The Separation of Economics from Virtue: A Historical-Conceptual Introduction," in *Economics and the Virtues: Building a New Moral Foundation,* Jennifer A. Baker and Mark D. White, eds. (New York, NY: Oxford University Press, 2016), 141.

[25] Thomas Kuhn, *The Structure of Scientific Revolutions,* 3rd ed. (Chicago, IL: University of Chicago Press, 1996), 160–161; see also 10–15.

[26] Popper is, in fact, quite clear about this. See Karl R. Popper, "Normal Science and its Dangers," in *Criticism and the Growth of Knowledge: Proceedings of the International Colloquium in the Philosophy of Science, London, 1965, Volume 4,* Imre Lakatos and Alan Musgrave, eds. (New York, NY: Cambridge University Press, 1970), 51.

[27] For a subtle analysis, see Liam Kofi Bright, "On Fraud," *Philosophical Studies* 174, no. 2 (2017): 291–310.

Commentary on Lecture 3

WHAT ABOUT TRUST? COMMUNICATION AND PUBLIC CONTROVERSIES ABOUT SCIENCE

Daniel Steel

Introduction

In her lecture "Science Communication: Beyond the Deficit Model," Douglas is concerned with public controversies about science in which some significant subgroups do not accept a scientific consensus on a topic, and reject the measures this consensus is taken to support. Examples fitting this pattern include anthropogenic climate change denial, anti-vaccination movements, and creationism. Douglas argues that common approaches to such situations either rest on false assumptions (e.g., the deficit model) or fail to suggest productive solutions (e.g., motivated cognition). Therefore, Douglas aims to provide a positive proposal for how to address politically charged public controversies such as these related to science.

In commenting on Douglas' proposals, I would like to begin by noting points of agreement. I agree with her criticisms of the deficit model. I also agree that empirical investigations of the role of motivated cognition in contentious topics do not indicate solutions. Moreover, I agree that more explicit recognition and discussion of the role of values can sometimes be helpful for communication on scientific issues. Given this, I focus here on Douglas' positive proposal, according to which conflicting sides in a controversy should specify what evidence would be sufficient to change their minds and why. The idea is that such a starting point can then be a basis for policy proposals that might be acceptable to both sides, or lead to the provision of new evidence that satisfies all parties to the dispute.

While I find much to admire in Douglas' proposal, I offer two critical observations intended in a constructive vein. The first is that the intended context of Douglas' proposal is not specified. Is it intended to apply to open, disorganized, and unstructured debates being carried out in the court of public opinion, encompassing venues such as popular news outlets, blog posts, and social media alongside more traditional modes of scientific communication? Or is it a model for a deliberative exercise in which representatives of stakeholder groups, including scientists and members of the public, are recruited to engage with one another in an organized and structured format? Without knowing the answers to such questions, it is difficult to assess what the proposal might reasonably hope to achieve or by what standards it should be evaluated.

The second concern is that the proposal faces difficulties relating to breakdowns of trust that pervade the examples of scientific controversies that Douglas mentions. That is, Douglas' proposal recommends assuming that all those involved in the controversy are rational and take the evidence seriously. Yet in some cases there may be good reason to think this assumption is false. Moreover, assuming the reasonableness and integrity of others despite evidence to the contrary can make one vulnerable to manipulation. Thus, I suggest that discussion of substantive conditions under which trust is, and is not, justified should be a component of any approach to controversies relating to science.

Douglas' Model for Science Communication

Douglas proposes that science communication on controversial topics such as climate change denial or anti-vaccination should avoid the patronizing attitude suggested by the deficit model ("let me explain the science to you, then you'll change your mind") as well as *ad hominem* insinuations that skepticism about a scientific consensus is simply the result of motivated cognition ("you just think that because it helps you get along with your coworkers in Oklahoma"). Instead, she proposes that:

> *We can start with the presumption that everyone is acting rationally and is taking the evidence seriously. We can ask all parties (even our own – this is entirely reflexive) why they view the evidence as supporting their view. This should allow them to elucidate both the evidential and methodological basis for their view and the value basis, including why they view the evidence as sufficient. We can then ask all parties what evidence would change their minds. If what is at issue are value-based disagreements about evidential sufficiency (i.e., a proper indirect role for values regarding evidential sufficiency rather than an improper direct role), some hypothetical evidence should tip the balance to their opponents' position, thus illuminating the sufficiency standards others are using.*[1]

Douglas considers two primary paths forward at this point.

The first is that at least one party refuses to say what evidence would make them change their minds. Then Douglas suggests that it can be concluded that members of this group are driven by some extra-scientific motive, such as political ideology or profit, rather than some genuine concern about scientific uncertainty – in short, they are being dogmatic. In this case, the discussion ends and it is plain to all which side in the controversy is being unreasonable.

The second possibility is that both sides articulate their evidential thresholds; they state the evidence that, if procured, would cause them to relinquish their position. Two further, not necessarily mutually exclusive, possibilities present themselves at this stage. The first of these is to explore the value judgments that underlie the difference, and possibly try to craft a compromise policy solution that would satisfy both. As an example of this, Douglas suggests market-based approaches to climate change mitigation, such as cap and trade schemes or a carbon tax. The second is to seek evidence that would settle the dispute. If such evidence is acquired, and agreement is reached as a result, the problem is solved. But if one side persistently refuses to accept evidence that, to all appearances, satisfies its demands, that is an indicator of dogmatism. Thus, as in the case when one party to the dispute refuses to state what evidence would cause it to change its views, in this situation it will gradually become apparent to all who is being unreasonable.

Douglas' proposal, then, is to create a setting in which it is possible to rationally discuss differing value judgments that are relevant to disputes about what scientific evidence is sufficient on a controversial topic, while at the same time maintaining scientific integrity and not permitting values to substitute for evidence. In what follows, I take a closer, sometimes critical, look at this idea.

The Court of Public Opinion or Deliberative Democracy?

The first question to ask about this proposal is exactly what sort of process it is recommending. At several points, Douglas speaks of an unspecified "we": "We can start with the presumption," or "We can ask," and so on. But just who is being referred to here? Scientists who adhere to the scientific consensus view on the topic, philosophers of science or other academics who research this topic, or any interested person? And in what context is the imagined conversation occurring? As noted in the introduction, at least two possible interpretations suggest themselves. The conversation might be some open public discourse in the court of public opinion or the marketplace of ideas (choose your favorite metaphor). Or it might be an organized and structured deliberative exercise involving a carefully selected mini-public.[2] Unfortunately, Douglas does not clarify what context she has in mind.

Begin by considering the "court of public opinion" interpretation of the proposal. In this case, who is the "we" who presumes the good faith of others in the debate and asks all parties (including its own) to state the evidence that would make it change its mind? And to whom should such requests be directed? Several options can be imagined. Perhaps the "we" consists of those committed to the scientific consensus on the issue, who may be scientists, other academics, civil society groups, or representatives of government agencies. Or perhaps "we" stands for any individual interested in the issue and willing to devote time to discussing it. No matter how one answers these questions, several difficulties immediately arise, because the various "sides" in an unstructured public debate are rarely sharply delineated. As a result, it is likely to be unclear what the suite of contrasting positions are, who is associated with each, and who is authorized and obligated to speak for them.

For example, consider a discussion of the sort Douglas imagines on the topic of anthropogenic climate change. Many different climate scientists and organizations agreeing with the scientific consensus might issue public statements, yet these will not always coincide on matters crucial to Douglas' proposal, such as what evidence suffices for which claims and which actions should be taken given values and uncertainties. Similarly, if members of a scientific consensus group, following Douglas' proposal, wish to ask skeptics of anthropogenic climate change to state what evidence would change their views, they will have many individuals, organizations, and websites to choose from who may differ from one another in a variety of ways. Thus, a group of scientific consensus individuals may be uncertain of its ability to speak for other like-minded people, and is likely to have difficulty identifying an authoritative source to speak for the other "side" (assuming that it is in fact a relatively cohesive unit). And to make matters worse, representatives of each "side" may have such doubts about the other. That is, they may doubt whether a person or organization asking them to state for the record what evidence would change their views represents an important constituency to whom they have an obligation to reply. In such cases, failure of response is not necessarily an indicator of dogmatism. People or organizations that receive large quantities of correspondence cannot be expected to reply to them all in a detailed fashion. Moreover, even if they do reply, it is unclear to what extent they can be taken to speak for others who hold similar views.

In contrast, these issues are more manageable if Douglas' proposal is a model for an organized deliberation among a selected group of stakeholders or members of the public. In a setting of this sort, individuals can be chosen to represent salient opposed positions in the controversy, and organizers can impose ground rules assumed in the model, for instance, that participants treat one another respectfully, that they carefully consider questions from the other side(s) about what they would judge to be sufficient evidence for claims at issue, make it understood that a failure to respond would be taken to indicate dogmatism, and so on. Of course, this would only constitute a sort of internal validity: within the context of the deliberative exercise, the conditions assumed by Douglas' proposal could be approximated. Such a process might have value for those who participate in it, or as a source of advice for policymakers. But whether those interested in the issue but not directly involved in the deliberation would view it as legitimately representing their concerns and interests is another matter entirely.

Bringing Trust into the Picture

Another concern about Douglas' proposal stems from the mutual distrust that frequently arises among opposed camps on politically controversial topics. This issue is exacerbated by the fact that in some cases such distrust may be justified.

Consider the initial prescription of Douglas' proposal that one assume that all sides are operating in good faith, that they are rational and take the evidence seriously. What if this presupposition is not true, and indeed there is evidence to think that some vocal participants in the debate are unreasonable or lack integrity? What if, in fact, some people are not being rational (plausible in the vaccine case, I think), or are more interested in pushing claims that will promote their bottom line than in "taking the evidence seriously" (as seems likely for some skeptics of anthropogenic climate change)? For example, in the case of anti-vaccine movements, there is very good reason to think that some leading actors, such as Andrew Wakefield, have been dishonest, driven by ulterior motives, and have engaged in fraudulent research.[3] Similar accusations have been leveled at some leading anthropogenic climate change skeptics[4] and, indeed, Douglas has directed such charges at Fred Singer.[5] Assuming the good faith of individuals such as these would appear to simply be naïve, and make one susceptible to being duped or manipulated.

This difficulty is especially acute if Douglas' proposal is taken to be a model for a discussion being carried out in public venues such as talk radio, partisan websites, blogs, and social media. Such venues do not contain mechanisms for filtering out dishonest or abusive communication, can become dominated by individuals who promote extreme views, and can create a vicious cycle in which like-minded individuals reinforce one another's biases. In short, such discussions are hardly a setting in which a presumption of reasonableness and careful concern for evidence would be appropriate. Moreover, when there are good reasons to suspect that other sides in the debate are not operating in good faith, there can be good reasons to refuse to engage in dialogue, for example, by refusing to answer requests to state what evidence would change one's mind. In situations of poisoned trust, refusal to participate may derive from a justifiable suspicion that participation would only make one susceptible to being manipulated into inadvertently supporting objectives one strongly opposes. Moreover, these concerns can be relevant for deliberative exercises wherein organizers attempt to impose some measure of order and civility.

To illustrate, consider ill-fated efforts by the French government to conduct deliberative democracy forums on the topic of nanotechnology in 2009 and 2010.[6] This attempt at deliberative democracy quickly devolved into a fiasco due to the refusal of opponents and skeptics of nanotechnology to participate in the process. Gaillard describes the situation as follows:

> *Contrary to many deliberative experiments based on the careful choice of a panel of citizens, either randomly or in a representative way according to the diversity of social backgrounds, these debates were open to all. If the meetings indeed occurred in some cities, the debate was not held in many of them, individuals or organizations preventing them*

from happening by force. Some debates thus had to be cancelled due to these protests (with banners, shouting…). These dissenters were putting forward the following argument: joining the debate means recognizing the legitimacy of the debate and the authority, which in turns means accepting nanotechnology, the political outcome of the debate suspected in advance to be the endorsement of R&D in nanotechnology.[7]

Moreover, the dissenters' argument appears to have had some merit. As Gaillard explains, "At the very same time that the public debate on nanotechnology was being set up in France, calls for proposals in research were already being launched in this field by research institutions or funding agencies."[8] Consequently, the dissenters may have been justified in regarding the deliberative exercise as little more than window dressing intended to legitimize a decision that had already been made. In such circumstances, refusal to participate may be reasonable and cannot be taken as a genuine indicator of dogmatism or values overriding evidence.

In sum, distrust is common and sometimes justified in public controversies related to science, and that in turn calls into question Douglas' prescription that all participants assume opposing camps are reasonable and take the evidence seriously. In some cases, there may be good reasons to doubt this assumption, and when this is so, naïvely assuming the integrity of certain individuals or organizations may make one vulnerable to manipulation. As a result, an ability to discern which sources of information are worthy of trust, and which are not, is important for navigating controversial topics effectively. Moreover, distrust may arise from longstanding social divisions, such as inequalities or historical injustices, which can hardly be assumed away by an act of will. In such circumstances, the important question is whether objective conditions required for trust are present and commonly recognized.[9]

Conclusion

In spite of the above, I believe that Douglas' proposal contains valuable insights, but for something more modest than a resolution of longstanding science-related controversies. Specifically, I think her proposal has considerable interest as an outline for a deliberative exercise involving stakeholders or members of the public. Within such a context, some of the concerns raised above can be mitigated by not including individuals with a track record of dishonesty or abusive behavior, and by insisting that rules for respectful and considered communication be followed. But such approaches, while potentially valuable, are modest, insofar as they do not promise to resolve the sorts of controversies—climate change denial, anti-vaccination, etc.—that Douglas proposes to address. There is no assurance that the public at large would accept their results, in part because deliberative exercises are unlikely to change the broader social forces that sustain distrust. I conclude, therefore, with the constructive suggestion that Douglas' proposal might benefit from greater attention to the role of trust and distrust in public controversies related to science.

Notes

[1] Heather Douglas, this volume, 134–135.

[2] These are not the only possible interpretations. For example, perhaps Douglas' proposal is a model for a learning exercise that could be carried out in a classroom.

[3] Dennis K. Flaherty, "The Vaccine-Autism Connection: A Public Health Crisis Caused by Unethical Medical Practices and Fraudulent Science," *Annals of Pharmacotherapy* 45, no. 10 (2011): 1302–1304.

[4] Naomi Oreskes and Erik Conway, *Merchants of Doubt: How a Handful of Scientists Obscured the Truth on Issues from Tobacco*

Smoke to Global Warming (New York, NY: Bloomsbury Press, 2010).

[5] Heather Douglas, "Bullshit at the Interface of Science and Policy: Global Warming, Toxic Substances and other Pesky Problems," in *Bullshit and Philosophy,* Gary Hardcastle and George Reisch, eds. (Chicago, IL: Open Court, 2006), 213.

[6] Maxence Gaillard, "The Governance of 'Well-Ordered Science,' from Ideal Conversation to Public Debate," *Theoria* 77 (2013): 245–256.

[7] Ibid., 252.

[8] Ibid., 253.

[9] Heidi Grasswick, "Scientific and Lay Communities: Earning Epistemic Trust through Knowledge Sharing," *Synthese* 177 (2010): 387–409; Kyle Whyte and Robert Crease, "Trust, Expertise, and the Philosophy of Science," *Synthese* 177 (2010): 411–425; Torsten Whilholt, "Epistemic Trust in Science," *British Journal for the Philosophy of Science* 64, no. 2 (2012): 233–253.

ABOUT THE CONTRIBUTORS

Heather Douglas

Heather Douglas is an associate professor in the Department of Philosophy at Michigan State University. She received her PhD from the History and Philosophy of Science Department at the University of Pittsburgh in 1998, and has held tenure-line positions since then at the University of Puget Sound, the University of Tennessee, and the University of Waterloo. She is the author of *Science, Policy, and the Value-Free Ideal* (2009), as well as numerous articles on values in science, the moral responsibilities of scientists, and the role of science in democratic societies. Her work has been supported by the National Science Foundation. In 2016, she was named an AAAS fellow.

Matthew J. Brown

Matthew J. Brown is director of the Center for Values in Medicine, Science, and Technology, Program Head, and professor of philosophy and history of ideas at the University of Texas at Dallas. He works in philosophy of science, science and technology studies, history of philosophy, and cognitive science. His monograph, *Science and Moral Imagination: A New Ideal for Values in Science* (University of Pittsburgh Press), explores the role of values in science and the scientific basis of values from a broadly pragmatist perspective. Brown received his PhD in Philosophy from UC San Diego. See more at www.matthewjbrown.net.

Sir Peter Gluckman

Sir Peter Gluckman is chair of the International Network of Government Science Advice (INGSA) and president-elect of the International Science Council (ISC). He heads Koi Tū: the Centre for Informed Futures in the University of Auckland. From 2009 to 2018, he was first chief science advisor to the prime ministers of New Zealand. Trained as a pediatrician and biomedical scientist, he holds a Distinguished University Professorship in the University of Auckland. He has received the highest civilian and academic honors in New Zealand and published over 700 scientific papers and written extensively on the relationship between evidence and policy and on science diplomacy.

Arthur C. Petersen

Arthur C. Petersen is a professor of science, technology and public policy in the Department of Science, Technology, Engineering and Public Policy (STEaPP) at University College London (UCL). He joined UCL STEaPP full time in September 2014 after more than 13 years' work as scientific adviser on environment and infrastructure policy within the Dutch government. He is editor of *Zygon: Journal of Religion and Science.* Most of his research involves dealing with uncertainty.

Kristina Rolin

Kristina Rolin is a university lecturer in research ethics at Tampere University. Her areas of research are philosophy of science and social science, social epistemology, and feminist epistemology and philosophy of science. She is interested in diversity in science, the role of trust and values in science, collective knowledge, and objectivity.

Eric Schliesser

Eric Schliesser is a professor of political science at the University of Amsterdam and Visiting Scholar at Chapman University's Smith Center for Political Economy and Philosophy. His research encompasses a variety of themes, ranging from economic statistics in classical Babylon, the history of the natural sciences, and forgotten 18th century feminists (both male and female) of political theory, and the history of political theory and the assumptions used in mathematical economics. He is the author of *Adam Smith: Systematic Philosopher and Public Thinker* (Oxford University Press) and keeps a daily blog called *Digressionsnimpressions*.

Daniel Steel

Daniel Steel is an associate professor at the University of British Columbia in the Centre for Applied Ethics in the School of Population and Public Health. His research interests lie in ethical and epistemic issues that arise at the crossroads of science and public policy. He is the author of *Philosophy and the Precautionary Principle: Science, Evidence, and Environmental Policy* (Cambridge University Press).

Torsten Wilholt

Torsten Wilholt is a professor of philosophy and history of the natural sciences at Leibniz Universität Hannover. He has worked and published on the social epistemology of science, the philosophy of applied science, the political philosophy of science and philosophy of mathematics. He studied philosophy, mathematics and history of science at Göttingen, Berlin, and Bielefeld, where he received both his PhD and his Habilitation in Philosophy. At Hannover, he is a cofounder of the interdisciplinary Leibniz Center for Science and Society as well as chairperson of "Integrating

Ethics and Epistemology of Scientific Research," a joint graduate program of the universities of Hannover and Bielefeld.

Made in the USA
Monee, IL
04 November 2026

67375584R00108